SRA

MATH SKILLBUILDER

yellow

SRA

Columbus, OH

Photo Credits
24 Juice Images/Alamy, **58** ©moodboard/Alamy, **74** ©DreamPictures/Shannon Faulk/Blend Images LLC, **176** Jupiterimages/Comstock Images/Getty Images, **196** (l to r)Photodisc Collection/Getty Images, (2)cgbaldauf/E-plus/Getty Images, (3)Daniel Parent/E-plus/Getty Images

SRAonline.com

SRA

Printed in the United States of America.

Send all inquiries to this address:
SRA/McGraw-Hill
4400 Easton Commons
Columbus, OH 43219

ISBN: 978-0-07-618620-4
MHID: 0-07-618620-2

8 9 10 11 12 13 QVS 20 19 18 17 16

Contents

NAME ____________________

YELLOW BOOK PRETESTS
Readiness Check

Add.

	a	b	c	d	e
1.	42 + 6	30 +40	51 +16	94 +8	66 +85
2.	743 +68	635 +245	1837 +4406	465 372 +56	6073 4125 5678 +1304

Subtract.

3.	86 − 4	70 −20	56 −36	40 −13	632 −57
4.	657 −305	734 −256	4016 −639	7524 −3258	26400 −9368

Multiply.

5.	50 ×9	41 ×6	39 ×8	634 × 7	2536 × 4
6.	46 ×15	237 ×68	573 ×458	907 ×216	8273 × 65

Divide.

	a	b	c	d
7.	$7\overline{)56}$	$5\overline{)35}$	$6\overline{)42}$	$8\overline{)64}$
8.	$9\overline{)56}$	$7\overline{)95}$	$6\overline{)309}$	$4\overline{)784}$
9.	$4\overline{)1652}$	$8\overline{)3249}$	$5\overline{)7625}$	$7\overline{)8617}$

Complete the following.

10.	$0.65 +0.37	$6.78 +8.69	$0.24 × 7	$5.60 × 37

	a	b
11.	7 cm = ________ mm	8 qt = ________ gal
12.	6 kg = ________ g	8 c = ________ pt
13.	9 km = ________ m	36 in. = ________ ft
14.	5 kL = ________ liters	9 ft = ________ yd

NAME ______________________________

YELLOW BOOK PRETESTS
Addition Facts (Pretest 1)

	a	b	c	d	e	f	g	h
1.	2 +2	5 +3	1 +1	8 +4	7 +1	0 +4	6 +3	3 +2
2.	7 +2	4 +1	6 +4	8 +0	4 +8	3 +7	4 +9	9 +3
3.	6 +5	0 +0	3 +8	8 +3	9 +5	6 +2	9 +0	2 +9
4.	4 +2	1 +3	5 +2	0 +2	9 +4	3 +6	8 +2	4 +7
5.	5 +4	8 +5	6 +6	2 +3	7 +3	2 +1	0 +7	1 +4
6.	9 +6	4 +3	7 +4	3 +5	5 +1	5 +5	8 +9	8 +1
7.	2 +4	9 +2	5 +9	9 +9	6 +7	2 +8	7 +5	4 +6
8.	8 +6	3 +4	2 +5	5 +6	7 +0	7 +6	9 +8	2 +7
9.	7 +9	4 +4	9 +1	6 +8	7 +8	5 +7	3 +9	8 +8
10.	6 +9	9 +7	3 +3	5 +8	2 +6	8 +7	4 +5	7 +7

NAME ______________________

YELLOW BOOK PRETESTS
Addition Facts (Pretest 2)

	a	*b*	*c*	*d*	*e*	*f*	*g*	*h*
1.	5 +2	1 +3	6 +6	0 +5	4 +0	3 +2	8 +3	2 +4
2.	3 +3	7 +1	2 +1	5 +8	7 +9	4 +4	6 +0	3 +6
3.	4 +3	0 +8	9 +2	7 +8	9 +9	0 +3	6 +7	4 +9
4.	5 +7	8 +4	3 +7	4 +8	6 +2	2 +2	5 +9	1 +1
5.	8 +9	1 +5	0 +0	4 +2	6 +8	8 +5	6 +5	4 +7
6.	2 +5	9 +7	5 +6	8 +2	3 +5	2 +8	9 +3	7 +2
7.	1 +6	6 +3	3 +8	9 +5	5 +5	7 +7	4 +5	2 +9
8.	6 +9	4 +6	8 +7	2 +6	1 +8	9 +8	7 +3	5 +4
9.	7 +5	8 +0	3 +4	7 +4	9 +6	4 +1	0 +9	8 +6
10.	2 +7	8 +8	5 +3	9 +4	3 +9	6 +4	7 +6	1 +9

NAME ______________________

YELLOW BOOK PRETESTS
Subtraction Facts (Pretest 1)

	a	b	c	d	e	f	g	h
1.	6 −5	14 −6	7 −2	11 −8	9 −4	10 −7	7 −6	14 −9
2.	9 −7	12 −5	7 −5	17 −8	8 −8	11 −4	6 −3	12 −8
3.	6 −6	10 −8	9 −2	11 −5	6 −2	15 −9	8 −6	13 −6
4.	11 −2	16 −7	7 −3	12 −9	5 −5	11 −7	8 −3	12 −4
5.	10 −1	15 −6	8 −7	10 −6	7 −1	10 −3	5 −2	13 −9
6.	9 −6	17 −9	2 −1	12 −3	8 −4	10 −5	9 −8	16 −8
7.	7 −7	15 −8	4 −2	16 −9	1 −1	13 −5	3 −1	10 −9
8.	9 −5	13 −4	8 −0	15 −7	9 −1	12 −6	7 −0	13 −8
9.	11 −3	12 −7	8 −1	11 −6	9 −0	11 −9	8 −5	14 −7
10.	18 −9	10 −4	9 −9	13 −7	10 −2	14 −8	9 −3	14 −5

YELLOW BOOK PRETESTS
Subtraction Facts (Pretest 2)

	a	b	c	d	e	f	g	h
1.	10 − 6	9 − 1	11 − 3	8 − 2	12 − 3	5 − 0	10 − 1	9 − 5
2.	11 − 7	4 − 0	12 − 5	10 − 9	11 − 4	9 − 8	13 − 4	8 − 8
3.	12 − 6	9 − 7	10 − 2	5 − 5	12 − 8	9 − 3	11 − 6	6 − 0
4.	14 − 5	11 − 9	12 − 4	9 − 6	11 − 5	8 − 1	13 − 7	7 − 3
5.	12 − 7	18 − 9	13 − 6	6 − 6	14 − 7	7 − 0	11 − 2	8 − 6
6.	10 − 3	9 − 2	10 − 5	5 − 1	13 − 8	6 − 5	17 − 8	7 − 1
7.	11 − 8	8 − 3	12 − 9	8 − 0	16 − 8	8 − 7	10 − 4	13 − 9
8.	10 − 7	7 − 5	15 − 8	6 − 2	14 − 6	14 − 9	15 − 7	9 − 4
9.	17 − 9	6 − 1	9 − 9	8 − 4	15 − 9	7 − 2	14 − 8	8 − 5
10.	13 − 5	9 − 0	16 − 9	7 − 6	16 − 7	10 − 8	15 − 6	7 − 7

NAME ____________________

YELLOW BOOK PRETESTS
Multiplication Facts (Pretest 1)

	a	*b*	*c*	*d*	*e*	*f*	*g*	*h*
1.	9 ×1	4 ×2	3 ×7	1 ×4	9 ×3	5 ×9	8 ×8	2 ×6
2.	4 ×9	7 ×7	2 ×0	6 ×5	3 ×6	1 ×2	7 ×6	6 ×4
3.	3 ×4	0 ×0	6 ×6	7 ×5	4 ×8	8 ×7	5 ×3	4 ×0
4.	6 ×0	8 ×9	3 ×8	8 ×0	1 ×8	7 ×8	4 ×7	8 ×6
5.	2 ×5	5 ×4	7 ×4	0 ×3	9 ×4	2 ×2	9 ×9	1 ×1
6.	8 ×1	1 ×5	6 ×7	7 ×9	3 ×1	8 ×5	3 ×5	6 ×3
7.	4 ×3	6 ×1	5 ×5	6 ×8	4 ×6	9 ×8	1 ×0	7 ×1
8.	6 ×9	0 ×5	8 ×4	3 ×3	7 ×0	2 ×8	9 ×7	3 ×9
9.	5 ×6	9 ×5	1 ×9	5 ×2	5 ×7	9 ×2	4 ×5	7 ×2
10.	3 ×2	6 ×2	8 ×3	4 ×4	0 ×9	7 ×3	5 ×8	9 ×6

NAME ____________________

YELLOW BOOK PRETESTS

Multiplication Facts (Pretest 2)

	a	*b*	*c*	*d*	*e*	*f*	*g*	*h*
1.	5 × 5	4 × 3	1 × 1	7 × 3	6 × 2	0 × 7	8 × 1	3 × 4
2.	6 × 3	9 × 7	2 × 1	3 × 3	8 × 2	4 × 1	9 × 5	2 × 8
3.	2 × 9	0 × 8	5 × 6	1 × 3	5 × 2	2 × 2	0 × 0	9 × 3
4.	3 × 5	4 × 4	8 × 3	7 × 2	0 × 1	4 × 9	8 × 8	3 × 0
5.	9 × 2	6 × 4	0 × 2	9 × 4	6 × 5	3 × 8	2 × 3	5 × 0
6.	5 × 1	7 × 4	8 × 4	2 × 4	9 × 6	5 × 7	7 × 9	8 × 7
7.	9 × 1	1 × 7	7 × 5	0 × 4	4 × 5	9 × 9	5 × 8	4 × 8
8.	6 × 8	8 × 5	5 × 9	6 × 6	7 × 7	0 × 6	5 × 3	3 × 9
9.	2 × 5	3 × 6	9 × 8	1 × 6	8 × 6	4 × 7	7 × 8	6 × 9
10.	5 × 4	7 × 6	8 × 9	4 × 6	9 × 0	6 × 7	3 × 7	2 × 7

NAME ______________________

YELLOW BOOK PRETESTS
Division Facts (Pretest 1)

	a	b	c	d	e	f	g
1.	2)6	9)18	3)15	6)18	1)3	4)12	5)45
2.	5)35	4)8	7)0	1)7	4)36	9)27	8)16
3.	2)8	6)24	9)36	3)18	4)16	7)7	3)12
4.	8)0	9)9	2)10	5)40	2)4	8)24	6)54
5.	2)2	6)0	4)32	3)21	9)45	3)9	7)14
6.	7)63	1)9	9)0	8)32	6)48	5)0	2)14
7.	5)30	4)28	7)56	2)12	8)72	1)5	9)54
8.	3)0	6)42	3)24	7)21	4)4	6)12	2)0
9.	7)28	8)40	5)25	7)49	5)5	9)63	8)64
10.	4)20	6)6	4)0	6)36	2)16	5)10	3)3
11.	1)8	5)20	4)24	9)72	8)56	7)42	3)27
12.	8)48	9)81	7)35	3)6	5)15	2)18	6)30

NAME ______________________

YELLOW BOOK PRETESTS
Division Facts (Pretest 2)

	a	b	c	d	e	f	g
1.	$3\overline{)18}$	$5\overline{)35}$	$4\overline{)4}$	$1\overline{)9}$	$7\overline{)0}$	$2\overline{)18}$	$4\overline{)36}$
2.	$6\overline{)54}$	$7\overline{)14}$	$2\overline{)16}$	$5\overline{)40}$	$4\overline{)8}$	$6\overline{)42}$	$7\overline{)63}$
3.	$1\overline{)0}$	$8\overline{)24}$	$4\overline{)32}$	$7\overline{)21}$	$1\overline{)6}$	$5\overline{)45}$	$3\overline{)0}$
4.	$5\overline{)30}$	$2\overline{)14}$	$6\overline{)48}$	$3\overline{)21}$	$7\overline{)28}$	$8\overline{)16}$	$9\overline{)9}$
5.	$3\overline{)15}$	$9\overline{)0}$	$1\overline{)5}$	$9\overline{)18}$	$3\overline{)6}$	$6\overline{)12}$	$8\overline{)40}$
6.	$7\overline{)35}$	$1\overline{)4}$	$8\overline{)48}$	$4\overline{)12}$	$8\overline{)8}$	$3\overline{)24}$	$5\overline{)0}$
7.	$2\overline{)12}$	$9\overline{)45}$	$4\overline{)0}$	$4\overline{)28}$	$1\overline{)3}$	$9\overline{)27}$	$6\overline{)36}$
8.	$4\overline{)24}$	$5\overline{)25}$	$2\overline{)10}$	$9\overline{)72}$	$5\overline{)10}$	$1\overline{)2}$	$8\overline{)56}$
9.	$6\overline{)24}$	$8\overline{)0}$	$7\overline{)49}$	$3\overline{)9}$	$4\overline{)20}$	$7\overline{)56}$	$2\overline{)0}$
10.	$3\overline{)12}$	$9\overline{)81}$	$1\overline{)1}$	$6\overline{)18}$	$5\overline{)15}$	$2\overline{)4}$	$9\overline{)54}$
11.	$6\overline{)6}$	$5\overline{)20}$	$6\overline{)30}$	$9\overline{)36}$	$2\overline{)8}$	$8\overline{)64}$	$3\overline{)27}$
12.	$8\overline{)32}$	$2\overline{)6}$	$8\overline{)72}$	$4\overline{)16}$	$6\overline{)0}$	$9\overline{)63}$	$7\overline{)42}$

YELLOW BOOK PRETESTS
Mixed Facts Pretest

Add, subtract, multiply, or divide. Watch the signs.

	a	*b*	*c*	*d*
1.	$\begin{array}{r}63\\+4\\\hline\end{array}$	$\begin{array}{r}49\\-8\\\hline\end{array}$	$\begin{array}{r}16\\\times 5\\\hline\end{array}$	$6\overline{)48}$
2.	$\begin{array}{r}56\\-13\\\hline\end{array}$	$9\overline{)37}$	$\begin{array}{r}85\\+7\\\hline\end{array}$	$\begin{array}{r}60\\\times 4\\\hline\end{array}$
3.	$\begin{array}{r}16\\\times 23\\\hline\end{array}$	$\begin{array}{r}35\\+42\\\hline\end{array}$	$7\overline{)53}$	$\begin{array}{r}26\\-18\\\hline\end{array}$
4.	$5\overline{)235}$	$\begin{array}{r}81\\\times 16\\\hline\end{array}$	$\begin{array}{r}639\\-18\\\hline\end{array}$	$\begin{array}{r}507\\+41\\\hline\end{array}$
5.	$\begin{array}{r}462\\+39\\\hline\end{array}$	$\begin{array}{r}483\\-57\\\hline\end{array}$	$\begin{array}{r}23\\\times 24\\\hline\end{array}$	$4\overline{)184}$
6.	$\begin{array}{r}506\\-273\\\hline\end{array}$	$\begin{array}{r}70\\\times 32\\\hline\end{array}$	$7\overline{)296}$	$\begin{array}{r}342\\+478\\\hline\end{array}$

Mixed Facts Pretest (continued)

	a	*b*	*c*	*d*
7.	$\begin{array}{r} 168 \\ \times 45 \\ \hline \end{array}$	$\begin{array}{r} 26 \\ 75 \\ 39 \\ +68 \\ \hline \end{array}$	$\begin{array}{r} 6392 \\ -418 \\ \hline \end{array}$	$14\overline{)296}$
8.	$32\overline{)278}$	$\begin{array}{r} 7320 \\ -465 \\ \hline \end{array}$	$\begin{array}{r} 913 \\ \times 69 \\ \hline \end{array}$	$\begin{array}{r} 365 \\ 491 \\ 872 \\ +516 \\ \hline \end{array}$
9.	$\begin{array}{r} 8070 \\ -416 \\ \hline \end{array}$	$\begin{array}{r} 784 \\ \times 123 \\ \hline \end{array}$	$\begin{array}{r} 6859 \\ 2306 \\ +7174 \\ \hline \end{array}$	$39\overline{)6842}$
10.	$\begin{array}{r} 6943 \\ 872 \\ 1064 \\ +5791 \\ \hline \end{array}$	$47\overline{)15960}$	$\begin{array}{r} 7000 \\ -5368 \\ \hline \end{array}$	$\begin{array}{r} 739 \\ \times 105 \\ \hline \end{array}$

PROBLEM SOLVING STRATEGIES

Multi-Step

Latoya and Eugene went to the store to buy a birthday present for their dad. They took $30.00 to spend. They bought a pack of golf balls that cost $7.50, a tie that cost $16.25, and a card that cost $1.75. How much money did Latoya and Eugene have left after buying these three items?

What operation do you do first?

They spent $25.50 on these three items.

What operation do you do next?

Latoya and Eugene have $4.50 left.

First find how much money Latoya and Eugene spent on the three items. To find this amount, **add.**

$$\begin{array}{r} \$7.50 \\ \$16.25 \\ +\$1.75 \\ \hline \$25.50 \end{array}$$

Next find how much money Latoya and Eugene have left. To find this amount, **subtract.**

$$\begin{array}{r} \$30.00 \\ -25.50 \\ \hline \$4.50 \end{array}$$

Solve each problem.

SHOW YOUR WORK

1. The Carsons drove from their home in Atlanta, Georgia, to Los Angeles, California, for a vacation. The total distance from Atlanta to Los Angeles is 1,946 miles. They drove 627 miles on Saturday and 715 miles on Sunday. They drove the remaining distance on Monday. How many miles did the Carsons drive on Monday?

What operation do you perform first? ____________________

The Carsons drove ________ miles on Monday.

2. At the souvenir store, Lee bought three key chains that cost $2.70 each and two T-shirts that cost $14.50 each. How much money did Lee spend at the souvenir store?

What operation do you perform first? ________________

Lee spent ________ at the souvenir store.

3. Antonio has a garden in his backyard that is 14 feet by 4 feet. Madeline has a garden in her backyard that is 9 feet by 6 feet. Whose garden has the larger perimeter? How many feet larger is the perimeter?

________________ garden perimeter is ________ larger.

PROBLEM SOLVING STRATEGIES

Draw a Picture

Camelia is having a party. Counting Camelia, there will be nine people at the party. Camelia is going to order pizza for the party. She wants to order enough pizza so that every person can have up to 3 slices of pizza. If the pizzas each have 8 slices, how many pizzas should Camelia order?

Camelia should order 4 pizzas.

Draw pizzas divided into eight slices. Place three 1's in three different slices to represent pizza for one person. Continue this through number 9. Count the number of pizzas needed for the party.

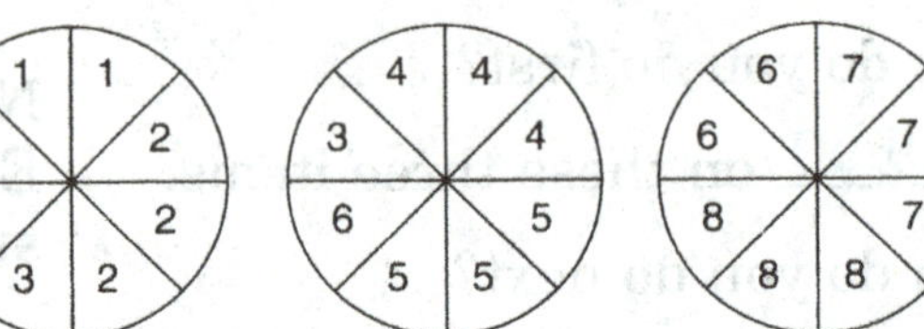
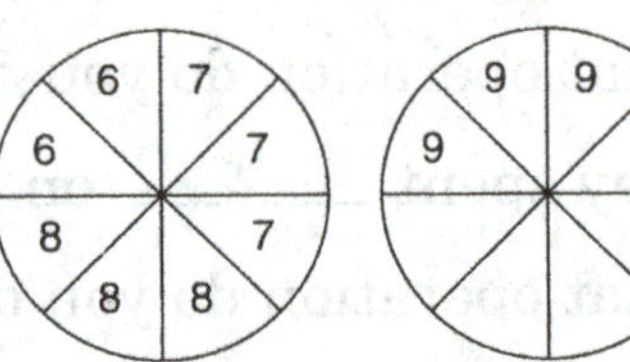

Solve each problem.

SHOW YOUR WORK

1. Tyra has a rectangular garden in her backyard that is 12 feet by 10 feet. She is going to put a fence around the garden. Between the fence and the garden there will be a 2-foot walkway. What is the perimeter of the fence around the garden and walkway?

   The perimeter of the fence will be ______ feet.

2. Marcus has a roll of plastic table covering that is 5 feet wide and 45 feet long. He is using it to cover picnic tables at a family picnic. All the picnic tables are 3 feet by 8 feet. What is the maximum number of picnic tables Marcus can cover with the roll of plastic table covering?

   Marcus can cover up to ______ picnic tables with the plastic table covering.

3. Aiyana is filling pages in her scrapbook. Each page of the scrapbook is 11 inches high and 10 inches wide. Each picture is 3 inches by 5 inches. What is the maximum number of pictures that she can fit on one page of her scrapbook?

   Aiyana can fit a maximum of ______ pictures on one page of her scrapbook.

NAME ______________________

Look for a Pattern

In January, Brenda put $10 in her savings account. In February, she put $15 in her savings account. In March, she put $20 in her savings account. If this pattern continues, how much money will Brenda put in her savings account in September?

Each month Brenda puts $5 more in her savings account than the previous month.

In September, Brenda will put $50 in her savings account.

Look for a pattern in the amounts Brenda put into her account in January, February, and March.

January		February		March
$10		$15		$20
	+$5		+$5	

Make a list of the amount she puts in her account each month until September.

Jan.	$10	June	$35
Feb.	$15	July	$40
Mar.	$20	Aug.	$45
Apr.	$25	Sept.	$50
May	$30		

Solve each problem.

SHOW YOUR WORK

1. Ken has tomato plants in his garden. He was able to pick his first tomatoes the first week of summer. That week he picked 4 tomatoes. The second week of summer he picked 5 tomatoes. The third week he picked 7 tomatoes. The fourth week he picked 10 tomatoes. If this pattern continues, how many tomatoes will Ken pick the eighth week of summer?

   In the eighth week of summer, Ken will pick ________ tomatoes.

2. Each birthday Rosa's mother measured her height. On Rosa's second birthday she was 35 inches tall. On her third birthday she was 38 inches tall. On her fourth birthday she was 41 inches tall. If this pattern continues, how tall will Rosa be on her ninth birthday?

   Rosa will be ________ inches tall on her ninth birthday.

PROBLEM SOLVING STRATEGIES

Guess and Check

Alexis has a total of \$24 in her wallet. She has a combination of \$1 bills and \$5 bills. She has a total of 12 bills. What combination of \$1 bills and \$5 bills does Alexis have in her wallet?

Alexis has __9__ \$1 bills in her wallet.

Alexis has __3__ \$5 bills in her wallet.

The number of bills must total 12.
Take a guess: 6 \$5 bills and 6 \$1 bills.

$\$5 \times 6 = \$30$
$\$1 \times 6 = \$6$
$\$30 + \$6 = \$36$ Too high

Try again: 2 \$5 bills and 10 \$1 bills.

$\$5 \times 2 = \$10$
$\$1 \times 10 = \$10$
$\$10 + \$10 = \$20$ Too low

Try again: 3 \$5 bills and 9 \$1 bills

$\$5 \times 3 = \$15$
$\$1 \times 9 = \$9$
$\$15 + \$9 = \$24$

Solve each problem.

1. Dontae has a collection of baseball cards and football cards. He has a total of 74 cards in all. He has 10 more baseball cards than football cards. How many of each type of card does Dontae have in his collection?

Dontae has ______ baseball cards in his collection.

Dontae has ______ football cards in his collection.

2. Katie is putting a fence around the rectangular garden in her backyard. She needs a total of 42 feet of fencing to go all the way around the garden. The length of the garden is twice as long as the width. What is the length and width of Katie's garden?

The length of Katie's garden is ______ feet.

The width of Katie's garden is ______ feet.

PROBLEM SOLVING STRATEGIES

Identify Missing Information

Over the summer Chayton went to the movie theater and watched 15 movies. He watched 4 action movies, 5 comedy movies, and the rest were scary movies or science fiction movies. How many scary movies did Chayton watch over the summer?

Not enough information.

Missing information: total number of science fiction movies Chayton watched

To find the number of scary movies, subtract the number of action, comedy, and science fiction movies from the total number of movies.

$$\begin{array}{rl} 15 & \\ -4 & \leftarrow \text{action movies} \\ \hline 11 & \\ -5 & \leftarrow \text{comedy movies} \\ \hline 6 & \\ - & \text{number of science fiction movies} \\ \hline \end{array}$$

Information on the number of science fiction movies is missing.

Solve each problem.

1. Olivia works Monday, Wednesday, Thursday, and Saturday at her part-time job. She works 4 hours on Monday and Thursday, and 5 hours on Saturday. How many hours does Olivia work each week at her part-time job?

____

Missing information: ______________________

____

2. During Saturday's football game, 135 soft drinks were sold at the concession stand. The soft drinks come in three different sizes: small, medium, and large. They sold 42 small soft drinks, and more large soft drinks than medium soft drinks. How many of each type of soft drink was sold on Saturday ?

____

Missing information: ______________________

____

____

NAME ______________________

PROBLEM SOLVING STRATEGIES

Make a Table

Bruce is following a recipe to make potato soup. For every three servings he needs two medium potatoes. If Bruce wants to make enough soup for 15 servings, how many medium potatoes does he need?

Bruce will need 10 medium potatoes for 15 servings of soup.

Make a table to determine the number of potatoes he will need for 15 servings.

Number of servings	Number of potatoes
3	2
6	4
9	6
12	8
15	10

Solve each problem.

SHOW YOUR WORK

1. Kashana is following a recipe to make pancakes. To make 6 pancakes she needs $\frac{1}{2}$ cup of pancake mix. How many cups of pancake mix will Kashana need to make 30 pancakes?

   Kashana will need ________ cups of pancake mix to make 30 pancakes.

2. A car rental company charged $45 the first day of renting a car and $25 every day after that. Jason rented a car for five days from the company. How much did it cost?

   It cost Jason ________ to rent a car for five days.

3. Bonita took $20 to the souvenir store. She wants to buy key chains to take home to her family and friends. Each key chain costs $3.00. What is the maximum number of key chains Bonita can buy with her money?

   Bonita can buy a maximum of ________ key chains with her money.

NAME ____________

PROBLEM SOLVING STRATEGIES

Make a List

PROBLEM SOLVING

The jungle tram ride at the amusement park departs every 25 minutes. The first one of the day leaves at 9:00 A.M. Jasmine and her friends want to get on the jungle tram ride that departs closest to 11:00 A.M. What time will Jasmine and her friends depart on the jungle tram ride?

Jasmine and her friends will depart on the jungle tram ride at 11:05 A.M..

Make a list of the departure times of the jungle tram ride.

9:00 A.M.
9:25 A.M.
9:50 A.M.
10:15 A.M.
10:40 A.M.
11:05 A.M. ← Closest to 11:00 A.M.

Solve each problem.

SHOW YOUR WORK

1. On Sunday, Lenno went for a walk and took a bike ride. He takes a walk every other day and goes for a bike ride every third day. What is the next day that Lenno will do both activities?

   The next day that Lenno will do both activities is ____________.

2. Molly is going to paint her kitchen and put up a wallpaper border. For the paint color, she is deciding between tan and cream. For the border, she is deciding between a flower pattern, a fruit pattern, or a basket weave pattern. How many different combinations of paint and wallpaper border can Molly choose from?

   Molly can choose from ________ different combinations of paint and wallpaper border.

3. For a weekend trip, Dylan packed a blue shirt, a white shirt, and a black shirt. He also packed a denim pair of shorts and a khaki pair of shorts. How many different outfits can Dylan choose from with the clothes he packed?

   Dylan can choose from ________ different outfits.

NAME ________________

PROBLEM SOLVING STRATEGIES

Work a Simpler Problem

Malcolm is building a wire fence around his swimming pool. The dimensions of the fence will be 21 feet 9 inches by 9 feet 3 inches. Malcolm went to the hardware store to buy the wire for the fence. A sales associate asked Malcolm for the perimeter of the fence.

The perimeter of Malcolm's new fence is ___62___ feet.

Adjust the dimensions so you can add simpler numbers. Take the 3 inches from 9 feet and add to the 9 inches to make 1 foot. Add the 1 foot to 21 feet. Use the dimensions 22 feet by 9 feet to find the perimeter.

To find the perimeter of the wire fence, add the lengths of all the sides.

$22 + 9 + 22 + 9 = 62$ feet

Solve each problem.

1. Mrs. Kelroy's classroom is 28 feet wide and 30 feet long. Mr. Price's classroom is 25 feet wide and 30 feet long. How many square feet larger is Mrs. Kelroy's classroom than Mr. Price's classroom?

Mrs. Kelroy's classroom is ________ square feet larger than Mr. Price's classroom.

2. Dean went shopping at a department store to buy some clothes. He took $50.00 with him. He wanted to buy a pair of shorts that cost $17.99, a T-shirt that cost $11.49, and cologne that cost $22.65. Did Dean have enough money to buy all three items?

Dean ________ have enough money to buy all three items.

PROBLEM SOLVING STRATEGIES

Work Backward

Jontell went shopping at the mall with some of her friends. She bought a hat that cost $9.45 and a sweatshirt that cost $20.64. After she bought these two items she had $4.91 left. How much money did Jontell take to the mall?

Jontell took $35.00 to the mall.

Work backward: Start with the amount left, $4.91, then add the amounts for the hat and the sweatshirt.

$4.91
20.64
+9.45
$35.00 ← beginning amount

Solve each problem.

SHOW YOUR WORK

1. Makoto, Sam, Julia, and Terrell are all cousins. Makoto is 3 years older than Sam. Sam is 2 years older than Julia. Julia is 2 years older than Terrell. Terrell is 5 years old. How old is Makoto?

Makoto is ________ years old.

2. On a math quiz, Michelle scored 7 points higher than Darnell. Darnell scored 3 points higher than Andrea. Andrea scored 1 point higher than Frankie. Frankie scored 81 points on the math quiz. What was Michelle's score on the math quiz?

Michelle scored a ________ on the math quiz.

3. Daniella stopped by a convenience store and bought a gallon of milk that cost $3.79 and a bag of chips that cost $2.24. The cashier gave her $3.97 back in change. How much money did Daniella give the cashier to pay for the milk and chips?

Daniella gave the cashier ________ to pay for the milk and chips.

PROBLEM SOLVING STRATEGIES

Estimation

Mrs. Whitmer stopped by the meat market to buy some meat for dinner. She wants to buy pork chops and roast. The pork chops cost $1.79 each and the roast costs $3.89 per pound.

She needs to buy 5 pork chops and 3 pounds of roast. About how much will she spend on the meat?

Use estimation to find the cost of each type of meat.

Mrs. Whitmer will spend about $22.00.

Since the pork chops cost $1.79 each, round to $2.00 each and multiply by 5.

$$\begin{array}{r} \$2.00 \\ \times\ 5 \\ \hline \$10.00 \end{array}$$

Since the roast costs $3.89 per pound, round to $4.00 per pound and multiply by 3.

$$\begin{array}{r} \$4.00 \\ \times\ 3 \\ \hline \$12.00 \end{array}$$

$10.00 + $12.00 = $22.00

Solve each problem.

SHOW YOUR WORK

1. At the bookstore, Juan wants to buy two magazines that cost $3.75 each and a book that costs $6.12. About how much will it cost for these three items?

It will cost about ________.

2. On Monday, 769 people visited the museum. On Tuesday, 524 people visited the museum. On Wednesday, 580 people visited the museum. About how many people visited the museum on Monday, Tuesday, and Wednesday?

About ________ people visited the museum.

3. For lunch, Bret is either going to buy soup and salad or a sandwich and chips. The soup is $1.79, a salad is $3.15, a sandwich is $4.95, and chips are $0.89. Which meal would be less expensive: soup and salad or sandwich and chips?

________________ would be a less expensive lunch.

NAME ____________________

CHAPTER 1 PRETEST

Addition and Subtraction (2-digit through 6-digit)

Add or subtract.

	a	*b*	*c*	*d*	*e*
1.	42 +26	37 +48	23 +95	76 +48	48 +39
2.	84 −23	75 −26	173 −92	165 −87	108 −39
3.	421 +357	832 +149	267 +138	521 +783	956 +287
4.	854 −321	783 −625	921 −570	1436 −349	1793 −875
5.	4235 +3796	6518 +4739	51672 +4318	52196 +38417	25186 +35821
6.	7659 −3847	8250 −6374	52169 −3057	42196 −38427	52105 −38156
7.	42 57 +38	34 27 +86	375 246 +381	6023 4034 +7012	73152 43081 +52165
8.	54 27 38 +46	731 208 319 +426	500 364 217 390 +324	8216 4315 2173 4081 +5216	70812 32181 31218 61408 +30802

Solve each problem.

1. How many points have been scored by both teams?

Kennedy has scored ________ points.

Clark has scored ________ points.

Both teams have scored ________ points.

2. Which team is ahead? By how many points are they ahead?

________ is ahead.

They are ahead by ________ points.

3. During the rest of the game Kennedy scored 10 more points and Clark scored 12 more points. Which team won the game? By how many points did they win?

The final score for Kennedy was ________.

The final score for Clark was ________.

________ won the game.

They won by ________ points.

1.

2.

3.

NAME ______________________________

Lesson 1 Addition Facts

CHAPTER 1

Add.

	a	*b*	*c*	*d*	*e*	*f*	*g*	*h*
1.	3 +6	7 +4	4 +3	5 +9	6 +1	6 +8	3 +7	9 +2
2.	8 +3	9 +5	8 +8	0 +5	6 +7	8 +2	7 +1	5 +7
3.	7 +9	4 +8	7 +2	3 +3	6 +6	4 +7	9 +1	6 +3
4.	6 +4	6 +9	2 +8	7 +7	8 +9	4 +2	4 +1	6 +2
5.	6 +5	4 +6	4 +4	3 +8	5 +2	7 +6	8 +0	3 +9
6.	5 +8	8 +1	3 +2	9 +3	7 +8	8 +6	7 +3	9 +8
7.	2 +7	9 +0	9 +6	2 +2	5 +4	8 +7	1 +9	7 +0
8.	3 +5	9 +7	5 +5	4 +9	9 +4	0 +7	3 +4	5 +6
9.	2 +5	8 +4	9 +9	7 +5	5 +3	2 +9	8 +5	2 +6

Lesson 2 Subtraction Facts

Subtract.

	a	b	c	d	e	f	g	h
1.	7 − 2	11 − 5	8 − 4	13 − 8	7 − 0	8 − 1	14 − 8	17 − 9
2.	8 − 3	11 − 4	9 − 2	14 − 9	12 − 8	11 − 7	4 − 0	10 − 1
3.	11 − 3	9 − 5	13 − 6	12 − 9	10 − 8	10 − 2	11 − 6	15 − 9
4.	15 − 8	11 − 2	10 − 3	16 − 9	14 − 7	14 − 5	12 − 4	7 − 3
5.	8 − 2	13 − 9	16 − 8	18 − 9	6 − 0	9 − 8	6 − 4	13 − 5
6.	8 − 7	15 − 7	10 − 4	7 − 5	12 − 6	13 − 4	13 − 7	6 − 2
7.	9 − 4	10 − 9	10 − 7	9 − 1	12 − 3	8 − 0	10 − 6	16 − 7
8.	14 − 6	10 − 5	11 − 9	6 − 5	5 − 2	9 − 7	15 − 6	9 − 6
9.	12 − 7	8 − 6	9 − 3	17 − 8	11 − 8	8 − 5	12 − 5	7 − 4

NAME ______________________

Lesson 3 Addition and Subtraction (2- and 3-digit)

	Add the ones. Rename.		Add the tens.
58 +89	8 +9 17	1 58 +89 7	1 58 +89 147

	Rename 146 as "1 hundred, 3 tens, and 16 ones." Then subtract the ones.	Rename 1 hundred and 3 tens as "13 tens." Then subtract the tens.
146 −87	3 16 1 4̸ 6̸ − 8 7 9	13 3 16 1̸ 4̸ 6̸ −8 7 5 9

Add.

	a	*b*	*c*	*d*	*e*	*f*
1.	23 +54	63 +25	72 +16	43 +54	26 +31	27 +42
2.	27 +35	47 +28	65 +26	31 +49	56 +28	39 +26
3.	47 +78	57 +86	32 +79	67 +84	36 +96	56 +47
4.	36 +27	45 +23	77 +77	63 +42	56 +24	35 +75

Subtract.

	a	*b*	*c*	*d*	*e*	*f*
5.	76 −24	37 −22	89 −63	75 −24	65 −31	49 −30
6.	95 −26	38 −19	52 −27	65 −48	91 −73	54 −27
7.	126 −37	143 −95	156 −88	172 −76	168 −99	153 −85

Lesson 3 Problem Solving

Solve each problem.

1. Sarah's father worked 36 hours one week and 47 hours the next week. How many hours did he work during these two weeks? He worked ________ hours the first week. He worked ________ hours the second week. During these two weeks, he worked a total of ________ hours.	**1.**
2. Seventy-six people live in Logan's apartment building. In Mike's apartment building, there are 85 people. How many more people live in Mike's building than in Logan's building? ________ people live in Mike's building. ________ people live in Logan's building. ________ more people live in Mike's building.	**2.**
3. In problem **2,** how many people live in both Logan's and Mike's apartment buildings? ________ people live in both buildings.	**3.**
4. There are 103 pages in Vera's new book. She has read 35 pages. How many pages does she have left to read? There are ________ pages in the book. She has read ________ pages. She has ________ pages left to read.	**4.**
5. Paula lives 53 kilometers from Darton. Ann lives 85 kilometers from Darton. How many kilometers closer to Darton does Paula live than Ann? Paula lives ________ kilometers closer.	**5.**

Lesson 4 Addition and Subtraction (3- and 4-digit)

Add from right to left.

carry 1 754 +587 1	carry 1 1 754 +587 41	carry 1 1 754 +587 1341

Subtract from right to left.

regroup: 1 3 4 1 → 3 11 (4 and 1 crossed out) 1341 −587 4	regroup: 2 13 11 (3, 4, 1 crossed out) 1341 −587 54	regroup: 12 13 11 (1, 3, 4, 1 crossed out) 1341 −587 754

Add.

	a	*b*	*c*	*d*	*e*	*f*
1.	314 +482	703 +192	542 +318	265 +429	553 +274	629 +280
2.	483 +702	546 +931	736 +279	653 +199	706 +539	582 +609
3.	813 +792	763 +762	423 +798	358 +759	816 +395	926 +178

Subtract.

	a	*b*	*c*	*d*	*e*	*f*
4.	784 −362	927 −405	542 −314	765 −238	926 −341	563 −281
5.	1732 −812	1574 −923	1764 −925	1345 −629	1542 −286	1637 −439
6.	1563 −678	1322 −733	1580 −687	1629 −243	1435 −162	1748 −358
7.	1984 −362	1864 −372	1250 −741	1608 −413	1500 −263	1542 −245

Lesson 4 Problem Solving

Answer each question.

1. The mileage reading on Mrs. Lee's car is 142. On Mr. Cook's, it is 319. How many more miles does Mr. Cook have on his car than Mrs. Lee?

Are you to add or subtract? ______________

How many more miles does Mr. Cook have on his car than Mrs. Lee? ________

1.

2. Carmen and Ava collect trading stamps. Carmen has 423 trading stamps and Ava has 519. How many stamps do both girls have?

Are you to add or subtract? ______________

How many stamps do both girls have? ______________

2.

3. Helen's family drove 975 miles on their vacation last year and 776 miles this year. How many miles did they travel during these two vacations?

Are you to add or subtract? ______________

How many miles did they travel during these two vacations? ________

3.

4. In problem **3**, how many more miles did they travel during the first year than the last?

Are you to add or subtract? ______________

How many more miles did they travel during the first year than the last? ________

4.

5. Tricia needs 293 more points to win a prize. It takes 1,500 points to win a prize. How many points does Tricia have now?

Are you to add or subtract? ______________

How many points does she have now? ________

5.

Lesson 5 Addition and Subtraction (4- and 5-digit)

Add.	21345 + 9462 = 30807			Subtract.	30807 − 9462 = 21345	
Check.	30807 − 9462 = 21345	These should be the same.		Check.	21345 + 9462 = 30807	These should be the same.

Add. Check each answer.

	a	b	c
1.	30821 + 4163	52964 + 3175	76487 + 5243
2.	42563 + 15786	15243 + 27561	36724 + 81409

Subtract. Check each answer.

	a	b	c
3.	72431 − 5316	92640 − 6741	61430 − 6429
4.	54061 − 6835	72413 − 6785	84205 − 5116

Lesson 5 Problem Solving

Solve each problem. Check each answer.

1. The space flight is expected to last 11,720 minutes. They are now 7,342 minutes into the flight. How many minutes remain?

   ________________ minutes remain in the flight.

2. In one year Mrs. Ching drove the company car 13,428 kilometers and her personal car 8,489 kilometers. How many kilometers did she drive both cars?

   She drove ________________ kilometers.

3. In problem **2,** how many fewer kilometers did she drive her personal car than the company car?

   She drove her personal car ________________ fewer kilometers.

4. The factory where Mr. Whitmal works produced 3,173 fewer parts this month than last. The factory produced 42,916 parts this month. How many parts did it produce last month?

   The factory produced ________________ parts last month.

5. Suppose the factory in problem **4** produced 3,173 more parts this month than last. How many parts would it have produced last month?

   ________________ parts would have been produced.

6. There are 86,400 seconds in a day. How many seconds are there in two days?

   There are ________________ seconds in two days.

7. During one month Jo Anne spent 14,400 minutes sleeping and 5,800 minutes eating. How much total time did she spend eating and sleeping?

   She spent ________________ minutes eating and sleeping.

1.

2.

3.

4.

5.

6.

7.

Lesson 6 Addition (3 or more numbers)

Add the ones.

3675
1406
3759
+6134

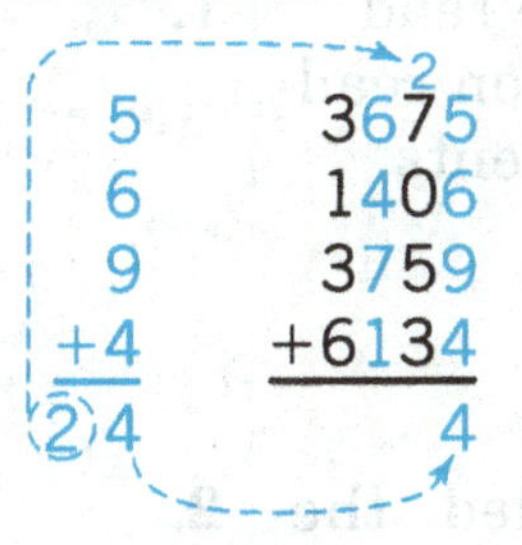

Follow the same pattern to add the tens, the hundreds, and so on.

1 1 2
3675
1406
3759
+6134
14974

Add.

	a	*b*	*c*	*d*	*e*
1.	453 216 +320	231 425 +317	242 375 +161	726 630 +712	542 416 +537
2.	6314 2145 +7634	2165 3420 +7015	8093 1246 +543	72193 83470 +21659	72165 45230 +3216
3.	325 463 179 +258	726 314 540 +829	7316 1425 7834 +2401	8216 7343 81692 +40830	92163 48517 73214 +82119
4.	730 460 273 892 +453	3829 1364 1274 429 +670	8213 4106 2300 4819 +2745	36000 72450 83192 62451 +31924	42165 30708 29115 40082 +31621
5.	542 365 421 300 460 +523	1628 329 1754 321 608 +2911	4216 53008 42134 2165 3008 +4000	52163 4218 316 5421 62190 +420	316 2143 126 52140 1230 +680

Lesson 6 Problem Solving

Solve each problem.

1. During the summer reading program, Faye read 752 pages. Barbara read 436 pages. Cameron read 521 pages. How many pages did these students read in all?

   They read ______ pages in all.

2. During September Joe Shedare traveled the following numbers of miles: 421; 308; 240; and 571. What was the total number of miles he traveled?

   He traveled a total of ______ miles.

3. Four astronauts have logged the following times in actual space travel: 4,216 minutes; 14,628 minutes; 3,153 minutes; and 22,117 minutes. How many minutes have all four astronauts logged in actual space travel?

   All four have logged ______ minutes in space.

4. The numbers of parts shipped to 6 cities were as follows: 317; 2,410; 32,415; 4,068; 321; and 5,218. How many parts were shipped in all?

   ______ parts were shipped.

5. A recent census gave the following populations: Adel, 4,321; Albany, 55,890; Alma, 3,515; Alto Park, 2,526; Americus, 13,472; and Ashburn, 3,291. What is the total population of these places?

   The total population is ______.

6. In an earlier census, the populations of the towns listed in problem **5** were 2,776; 31,155; 2,588; 1,195; 11,389; and 2,918, respectively. What was the total population then?

   Then the total population was ______.

7. In problem **5**, what is the total population of Adel, Albany, and Alto Park?

   The total population is ______.

1.
2.
3.
4.
5.
6.
7.

NAME ____________________

Lesson 7 Estimating Sums and Differences

Round each number to the highest place value the numbers have in common. Then add from right to left.

```
 382 ──→  400
+733 ──→ +700
         1100
```

Round each number to the highest place value they have in common. Then subtract from right to left.

```
27125 ──→ 27000
-8734 ──→ -9000
          18000
```

Estimate each sum or difference.

	a	b	c	d	e
1.	68 + 34	27 + 81	29 + 36	93 + 45	174 + 88
2.	47 − 26	94 − 58	66 − 57	82 − 34	373 − 89
3.	483 + 237	668 + 319	271 + 926	2325 + 760	827 + 958
4.	723 − 285	543 − 164	239 − 124	678 − 135	2923 − 1765
5.	4165 + 7584	1219 + 3470	9843 + 5194	32621 + 8319	22473 + 15704
6.	9261 − 3425	8498 − 2757	9032 − 3940	45462 − 24318	13852 − 9465
7.	561 + 807 + 534	3615 + 8124 + 4319	9273 + 12405 + 6793	29618 + 91128 + 73584	107374 + 42981 + 64563

Lesson 7 Problem Solving

Solve each problem. Use estimation.

1. On Monday, Carlos drove 273 miles. On Tuesday, he drove 129 miles. About how many more miles did Carlos drive on Monday than on Tuesday?

Are you to add or subtract? ____________

Carlos drove about ____________ more miles.

1.

2. Deena and Kayla both keep all their coins in a jar. Deena has 83 coins in her jar and Kayla has 76 coins in her jar. Together, about how many coins do Deena and Kayla have in their jars?

Are you to add or subtract? ____________

Deena and Kayla have about ____________ coins.

2.

3. At the arcade you get tickets by winning at several different games. After a year, Malcolm had saved 2,391 tickets and Brent had saved 3,608 tickets. If Malcolm and Brent combine their tickets to get one big prize, about how many tickets do they have?

Are you to add or subtract? ____________

Malcolm and Brent have about ____________ tickets.

3.

4. The Oakdale School auditorium holds 3,285 people. The Oakdale School gymnasium holds 1,340 people. About how many more people does the auditorium hold than the gymnasium?

Are you to add or subtract? ____________

The auditorium holds about ____________ more people.

4.

5. Jonica kept track of the number of pages she read each month over the summer. In June she read 319 pages. In July she read 185 pages. In August she read 227 pages. About how many pages did Jonica read during the summer months?

Are you to add or subtract? ____________

Jonica read about ____________ pages.

5.

CHAPTER 1 PRACTICE TEST

Addition and Subtraction (2-digit through 6-digit)

Add or subtract.

	a	b	c	d	e
1.	46 +32	423 +268	1829 +3573	7521 +3609	52163 +72845
2.	85 −32	564 −382	1936 −479	18312 −9264	10306 −2568
3.	32 26 +13	724 380 +465	295 327 168 +269	5534 1468 3137 +2950	42163 30820 21911 +60422
4.	7832 −1467	8309 −2654	13182 −4296	171234 −82169	102085 −36526

Solve each problem.

5. The following points were earned in a ticket-selling contest: Ashley, 2,320; Taylor, 1,564; Ali, 907; Lyn, 852; Marty, 775. What was the total number of points earned by Ashley and Ali?

Ashley earned ____________ points.

Ali earned ____________ points.

They earned a total of ____________ points.

6. In problem **5**, what was the total number of points earned by all five students?

They earned a total of ____________ points.

7. In problem **5**, how many more points did Taylor earn than Marty?

Taylor earned ____________ more points.

5.

6.

7.

CHAPTER 2 PRETEST

Multiplication (2-digit by 1-digit through 4-digit by 3-digit)

Multiply.

	a	*b*	*c*	*d*
1.	24 ×2	35 ×2	154 ×6	678 ×9
2.	31 ×23	82 ×18	45 ×51	87 ×39
3.	143 ×22	734 ×19	253 ×62	708 ×36
4.	321 ×123	432 ×621	507 ×143	821 ×105
5.	3126 ×422	4032 ×145	3124 ×712	8197 ×325

NAME ______________________

Lesson 1 Multiplication Facts

CHAPTER 2

Multiply.

	a	*b*	*c*	*d*	*e*	*f*	*g*	*h*
1.	4 ×2	8 ×2	7 ×2	2 ×2	6 ×2	5 ×2	3 ×2	1 ×2
2.	8 ×3	2 ×3	9 ×3	6 ×3	5 ×3	0 ×3	4 ×3	3 ×3
3.	2 ×4	1 ×4	6 ×4	8 ×4	7 ×4	3 ×4	9 ×4	4 ×4
4.	7 ×5	5 ×5	2 ×5	6 ×5	4 ×5	9 ×5	3 ×5	8 ×5
5.	6 ×6	2 ×6	9 ×6	3 ×6	1 ×6	7 ×6	5 ×6	8 ×6
6.	1 ×7	3 ×7	9 ×7	2 ×7	6 ×7	7 ×7	8 ×7	5 ×7
7.	5 ×8	1 ×8	7 ×8	2 ×8	9 ×8	6 ×8	3 ×8	8 ×8
8.	8 ×9	2 ×9	1 ×9	6 ×9	7 ×9	4 ×9	5 ×9	9 ×9
9.	3 ×0	8 ×0	0 ×0	1 ×0	6 ×1	2 ×1	9 ×1	7 ×1

Lesson 1 Problem Solving

Solve each problem.

1. There are six rows of desks in the office. Each row has eight desks. How many desks are in the office?

   There are ________ rows of desks.

   There are ________ desks in each row.

   There are ________ desks in all.

2. There are nine rows of trees. There are seven trees in each row. How many trees are there in all?

   There are ________ rows of trees.

   There are ________ trees in each row.

   There are ________ trees in all.

3. The people at the park were separated into teams of eight people each. Nine teams were formed. How many people were in the park?

   Each team had ________ people.

   There were ________ teams formed.

   There were ________ people in the park.

4. There are six people in each car. There were seven cars. How many people were there in all?

   There were ________ people in each car.

   There were ________ cars.

   There were ________ people in all.

5. How many cents would you need to buy eight 8-cent pencils?

   You would need ________ cents.

6. There are five oranges in each sack. How many oranges would there be in nine sacks?

   There would be ________ oranges in nine sacks.

1.

2.

3.

4.

5.

6.

NAME ______________________

Lesson 2 Multiplication (by 1-digit)

Multiply 3 ones by 5.

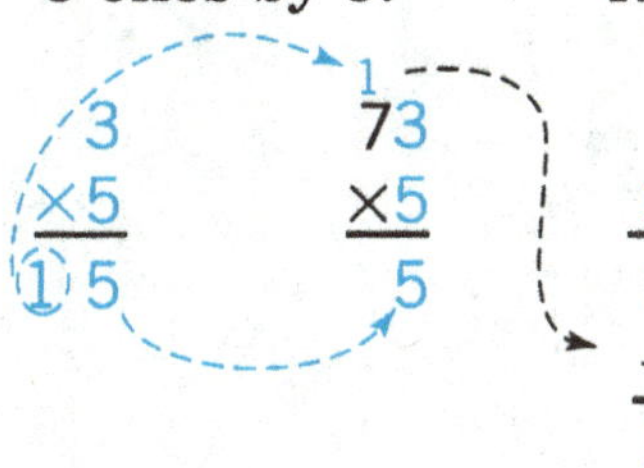

Multiply 7 tens by 5. Add the tens.

7 tens ×5 35 tens +1 ten 36 tens	$\overset{1}{7}3$ ×5 365

$3\overset{2}{2}7$ ×4 8	$3\overset{1}{2}\overset{2}{7}$ ×4 08	$\overset{1}{3}\overset{2}{2}7$ ×4 1308

Multiply.

	a	*b*	*c*	*d*	*e*	*f*
1.	32 ×2	21 ×3	42 ×2	132 ×2	213 ×3	421 ×2
2.	16 ×4	36 ×2	28 ×3	123 ×4	127 ×3	215 ×4
3.	73 ×3	42 ×4	81 ×5	352 ×2	172 ×4	263 ×3
4.	57 ×5	28 ×6	37 ×4	256 ×3	385 ×2	177 ×5
5.	28 ×6	47 ×8	39 ×5	426 ×7	358 ×6	234 ×5
6.	57 ×8	48 ×2	70 ×5	526 ×3	409 ×5	730 ×7
7.	72 ×9	95 ×7	81 ×8	629 ×8	801 ×7	658 ×9

Lesson 2 Problem Solving

Solve each problem.

1. Each club member works 3 hours each month. There are 32 members. What is the total number of hours worked each month by all club members?

There are ________ club members.

Each member works ________ hours.

The club members work ________ hours in all.

1.

2. Mrs. Robins drives 19 miles every working day. How many miles does she drive in a five-day workweek?

She drives ________ miles every working day.

She works ________ days a week.

She drives ________ miles in a five-day workweek.

2.

3. It takes 54 minutes to make one pillow. How long will it take to make 3 pillows?

It takes ________ minutes to make one pillow.

There are ________ pillows.

It takes ________ minutes to make 3 pillows.

3.

4. Each box weighs 121 kilograms. There are 4 boxes. What is the total weight of the 4 boxes?

Each box weighs ________ kilograms.

There are ________ boxes.

The total weight of the 4 boxes is ________ kilograms.

4.

5. There are 168 hours in a week. How many hours are there in 6 weeks?

There are ________ hours in 6 weeks.

6. There were 708 employees at work today. Each employee worked 8 hours. How many hours did these employees work?

________ hours were worked.

5.

6.

NAME ____________________

Lesson 3 Multiplication (by 2-digit)

41	41	56	56
×2	×20	×3	×30
82	820	168	1680

If 2 × 41 = 82, then 20 × 41 = ________.

If 3 × 56 = 168, then 30 × 56 = ________.

If 4 × 27 = 108, then 40 × 27 = ________.

Multiply 56 by 1.	Multiply 56 by 30.	
56	56	56
×31	×31	×31
56	56	56
	1680	1680 } Add.
		1736

CHAPTER 2

Multiply.

	a	*b*	*c*	*d*	*e*	*f*
1.	23 ×3	23 ×30	43 ×2	43 ×20	51 ×4	51 ×40
2.	37 ×4	37 ×40	54 ×6	54 ×60	73 ×9	73 ×90
3.	42 ×30	75 ×20	54 ×40	62 ×70	84 ×60	32 ×50

Multiply.

	a	*b*	*c*	*d*	*e*
4.	31 ×23	42 ×33	45 ×12	17 ×35	36 ×24
5.	54 ×26	37 ×41	28 ×16	38 ×73	46 ×28

Lesson 3 Problem Solving

Solve each problem.

1. There are 60 minutes in one hour. How many minutes are there in 24 hours?

   There are ________ minutes in 24 hours.

2. Forty-eight toy boats are packed in each box. How many boats are there in 16 boxes?

   There are ________ boats in 16 boxes.

3. Seventy-three new cars can be assembled in one hour. At that rate, how many cars could be assembled in 51 hours?

   ________ cars could be assembled in 51 hours.

4. A truck is hauling 36 bags of cement. Each bag weighs 94 pounds. How many pounds of cement are being hauled?

   ________ pounds of cement are being hauled.

5. To square a number means to multiply the number by itself. What is the square of 68?

   The square of 68 is ________.

6. Sixty-five books are packed in each box. How many books are there in 85 boxes?

   There are ________ books in 85 boxes.

7. Every classroom in Jan's school has at least 29 desks. There are 38 classrooms in all. What is the least number of desks in the school?

   There are at least ________ desks.

8. Some students came to the museum on 38 buses. There were 58 students on each bus. How many students came to the museum by bus?

   ________ students came by bus.

1.	2.
3.	4.
5.	6.
7.	8.

NAME ____________________

Lesson 4 Multiplication (by 2-digit)

	Multiply 351 by 7.	Multiply 351 by 20.	
351 ×27	351 × 27 2457	351 × 27 2457 7020	351 × 27 2457 7020 } Add. 9477

Multiply.

	a	b	c	d	e
1.	42 ×13	23 ×32	54 ×41	37 ×26	58 ×19
2.	58 ×72	27 ×36	40 ×55	27 ×27	39 ×42
3.	154 ×13	231 ×26	251 ×41	312 ×32	415 ×47
4.	365 ×27	426 ×13	715 ×26	302 ×43	756 ×29

Lesson 4 Problem Solving

Solve each problem.

1. A machine can produce 98 parts in one hour. How many parts could it produce in 72 hours?

   It could produce ________ parts in 72 hours.

2. Each new bus can carry 66 passengers. How many passengers can ride on 85 new buses?

   ________ passengers can ride on 85 buses.

3. A gross is twelve dozen or 144. The school ordered 21 gross of pencils. How many pencils were ordered?

   The school ordered ________ pencils.

4. How many hours are there in a year (365 days)?

   There are ________ hours in a year.

5. Each of 583 people worked a 40-hour week. How many hours of work was this?

   It was ________ hours of work.

6. The highway mileage between New York and Chicago is 840 miles. How many miles would a bus travel in making 68 one-way trips between New York and Chicago?

   The bus would travel ________ miles.

7. The airline distance between the cities in problem **6** is 713 miles. What is the least number of miles a plane would travel in making 57 one-way trips?

   The least number of miles would be ________ .

8. The rail mileage between Washington, D.C., and Chicago is 768 miles. How many miles would a train travel in making 52 one-way trips?

   It would travel ________ miles.

9. The airline distance between the cities in problem **8** is 597 miles. What is the least number of miles a plane would travel in making 45 one-way trips?

   The least number of miles would be ________.

1.	2.
3.	4.
5.	6.
7.	8.
9.	

NAME ______________________

Lesson 5 Multiplication (by 3-digit)

3254 ×2 = 6508	3254 ×20 = 65080	3254 ×200 = 650800	

If 2 × 3254 = 6508, then 20 × 3254 = __________.

If 2 × 3254 = 6508, then 200 × 3254 = __________.

```
  3254
 ×213
  9762   ——— 3 × 3254
 32540   ——— 10 × 3254
650800   ——— 200 × 3254
693102       Add.
```

CHAPTER 2

Multiply.

	a	*b*	*c*	*d*
1.	316 ×2	316 ×200	4281 ×3	4281 ×300
2.	416 ×213	375 ×291	408 ×316	219 ×503
3.	316 ×275	483 ×211	4231 ×213	3456 ×123
4.	2175 ×243	3216 ×208	3090 ×752	6613 ×342

Lesson 5 Problem Solving

Solve each problem.

1. Each crate the men unloaded weighed 342 pounds. They unloaded 212 crates. How many pounds did they unload?

   The men unloaded ____________ pounds.

2. The school cafeteria expects to serve 425 customers every day. At that rate, how many meals will be served if the cafeteria is open 175 days a year?

   ____________ meals will be served.

3. There are 168 hours in one week. How many hours are there in 260 weeks?

   There are ____________ hours in 260 weeks.

4. There are 3,600 seconds in one hour and 168 hours in one week. How many seconds are there in one week?

   There are ____________ seconds in one week.

5. A jet carrying 128 passengers flew 2,574 miles. How many passenger-miles (number of passengers times number of miles traveled) did it fly?

   It flew ____________ passenger-miles.

6. How many passenger-miles would be flown by the jet in problem **5** if it flew from Seattle to New Orleans, a distance of 2,098 miles?

   It would be ____________ passenger-miles.

7. A tank truck made 275 trips in a year. It hauled 5,950 gallons each trip. How many gallons did it haul that year?

   It hauled ____________ gallons.

8. Suppose the truck in problem **7** hauled 8,725 gallons each trip. How many gallons would it haul?

   It would haul ____________ gallons.

1.	2.
3.	4.
5.	6.
7.	8.

NAME ______________

Lesson 6 Estimating Products

Round 68 to 70. Then multiply from right to left.

68 → 70
×3   ×3
      210

Round each number to its highest place value.

625 → 600
×381 → ×400
       240000

Estimate each product.

	a	*b*	*c*	*d*	*e*
1.	84 × 7	43 × 6	87 × 3	66 × 9	91 × 5
2.	384 × 4	618 × 9	382 × 7	633 × 2	908 × 8
3.	28 × 37	39 × 75	76 × 18	48 × 37	63 × 47
4.	943 × 56	249 × 33	164 × 55	116 × 89	649 × 35
5.	329 × 607	261 × 329	892 × 219	740 × 273	819 × 464
6.	1862 × 97	6208 × 73	4935 × 48	1206 × 67	3496 × 13
7.	9815 × 264	4806 × 492	5930 × 228	2608 × 691	7393 × 535

Lesson 6 Problem Solving

Solve each problem. Use estimation.

1. Tyron read 18 pages each night for one week. About how many pages did he read that week? Tyron read about ________ pages that week.	**1.**
2. One type of airplane can carry up to 118 passengers. About how many passengers can five airplanes carry? Five airplanes can carry about ________ passengers.	**2.**
3. Kim watched three movies one weekend that were 96 minutes each. About how many minutes did Kim spend watching movies? Kim spent about ________ minutes watching movies one weekend.	**3.**
4. Elan has a paper route. He delivers 22 newspapers each hour. If it takes him 4 hours to deliver all the newspapers in his route, about how many newspapers does he deliver? Elan delivers about ________ newspapers on his paper route.	**4.**
5. A machine produces 367 gadgets in one hour. Each day the machine continually runs for 18 hours. About how many gadgets does the machine produce in one day? The machine produces about ________ gadgets in one day.	**5.**
6. Jasmine is on the cross-country team. Each day the team practices, the team members run 5 kilometers. If the team will have 32 practices this season, about how many kilometers will Jasmine run at practice this season? Jasmine will run about ________ kilometers at practice this season.	**6.**

NAME ______________________

CHAPTER 2 PRACTICE TEST

Multiplication (2-digit by 1-digit through 4-digit by 3-digit)

CHAPTER 2

Multiply.

	a	*b*	*c*	*d*
1.	31 × 3	25 × 3	276 × 6	583 × 7
2.	23 × 13	42 × 26	38 × 17	53 × 45
3.	123 × 31	425 × 70	563 × 25	837 × 85
4.	213 × 132	421 × 378	256 × 108	845 × 374
5.	1221 × 312	1456 × 173	1827 × 570	3456 × 732

NAME ______________________

CHAPTER 3 PRETEST
Division (2-digit through 4-digit by 1-digit)

CHAPTER 3

Divide.

	a	*b*	*c*	*d*
1.	$7\overline{)63}$	$6\overline{)54}$	$5\overline{)75}$	$4\overline{)92}$
2.	$4\overline{)136}$	$5\overline{)370}$	$3\overline{)471}$	$2\overline{)960}$
3.	$3\overline{)1539}$	$4\overline{)3672}$	$7\overline{)7105}$	$5\overline{)8605}$
4.	$4\overline{)87}$	$2\overline{)75}$	$3\overline{)86}$	$3\overline{)781}$
5.	$6\overline{)143}$	$4\overline{)9226}$	$2\overline{)1435}$	$5\overline{)6134}$

NAME ______________________

Lesson 1 Division Facts

$$\begin{array}{r} 9 \\ \times 5 \\ \hline 45 \end{array} \longrightarrow 5\overline{)45} \qquad \begin{array}{r} 9 \\ \times 5 \\ \hline 45 \end{array} \longrightarrow 9\overline{)45}$$

If $5 \times 9 = 45$, then $45 \div 5 = 9$ and $45 \div 9 = 5$.

CHAPTER 3

Divide.

	a	*b*	*c*	*d*	*e*	*f*
1.	$2\overline{)6}$	$3\overline{)9}$	$2\overline{)4}$	$2\overline{)8}$	$3\overline{)6}$	$4\overline{)8}$
2.	$1\overline{)5}$	$3\overline{)3}$	$6\overline{)0}$	$1\overline{)9}$	$2\overline{)2}$	$7\overline{)7}$
3.	$4\overline{)28}$	$6\overline{)42}$	$3\overline{)18}$	$6\overline{)36}$	$8\overline{)32}$	$2\overline{)14}$
4.	$2\overline{)10}$	$8\overline{)72}$	$7\overline{)42}$	$5\overline{)20}$	$3\overline{)15}$	$4\overline{)36}$
5.	$8\overline{)24}$	$2\overline{)18}$	$1\overline{)8}$	$4\overline{)32}$	$5\overline{)25}$	$9\overline{)81}$
6.	$7\overline{)35}$	$9\overline{)27}$	$6\overline{)24}$	$7\overline{)49}$	$8\overline{)48}$	$9\overline{)36}$
7.	$5\overline{)40}$	$3\overline{)24}$	$2\overline{)16}$	$6\overline{)48}$	$7\overline{)28}$	$9\overline{)54}$
8.	$5\overline{)15}$	$4\overline{)12}$	$2\overline{)12}$	$3\overline{)0}$	$6\overline{)54}$	$3\overline{)27}$
9.	$4\overline{)20}$	$8\overline{)56}$	$6\overline{)30}$	$4\overline{)24}$	$3\overline{)21}$	$5\overline{)30}$
10.	$8\overline{)16}$	$5\overline{)35}$	$4\overline{)16}$	$8\overline{)64}$	$9\overline{)63}$	$8\overline{)40}$

Lesson 1 Problem Solving

Solve each problem.

1. There are 18 chairs and 6 tables in the room. There are the same number of chairs at each table. How many chairs are at each table?

   There are ________ chairs.

   There are ________ tables.

   There are ________ chairs at each table.

2. Each box takes 3 minutes to fill. It took 18 minutes to fill all the boxes. How many boxes are there?

   It took ________ minutes to fill all the boxes.

   It takes ________ minutes to fill 1 box.

   There are ________ boxes.

3. Rob, Jose, Jay, Tom, Alex, and Jim share 6 sandwiches. How many sandwiches does each boy get?

   There are ________ sandwiches in all.

   The sandwiches are shared among ________ boys.

   Each boy gets ________ sandwich.

4. Bill and 8 friends each sold the same number of tickets. They sold 72 tickets in all. How many tickets were sold by each person?

   Each person sold ________ tickets.

5. Forty-eight oranges are in a crate. The oranges are to be put into bags of 6 each. How many bags can be filled?

   ________ bags can be filled.

6. Adam has a wire that is 42 inches long. He cuts the wire into 7-inch lengths. How many pieces of wire will he have?

   He will have ________ pieces of wire.

1.
2.
3.
4.
5.
6.

Lesson 2 Division (by 1-digit)

Study how to divide 738 by 3.

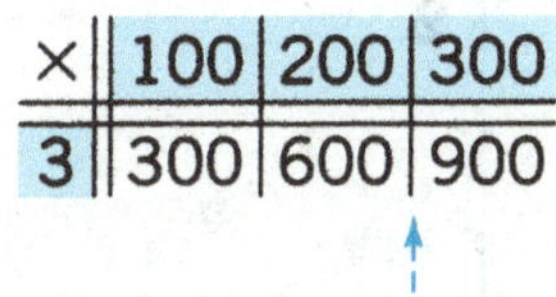

×	100	200	300
3	300	600	900

738 is between 600 and 900, so 738 ÷ 3 is between 200 and 300. The hundreds digit is 2.

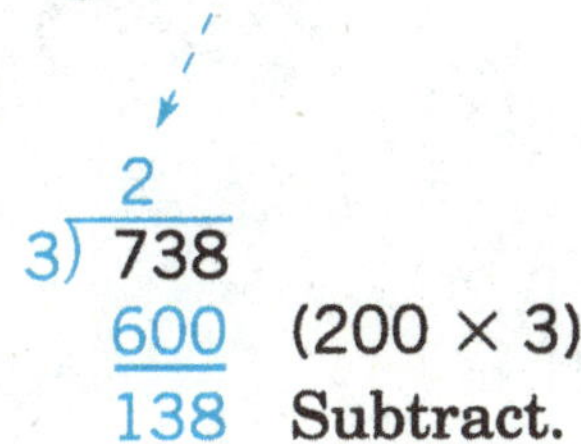

```
    2
3) 738
   600   (200 × 3)
   138   Subtract.
```

×	10	20	30	40	50
3	30	60	90	120	150

138 is between 120 and 150, so 138 ÷ 3 is between 40 and 50. The tens digit is 4.

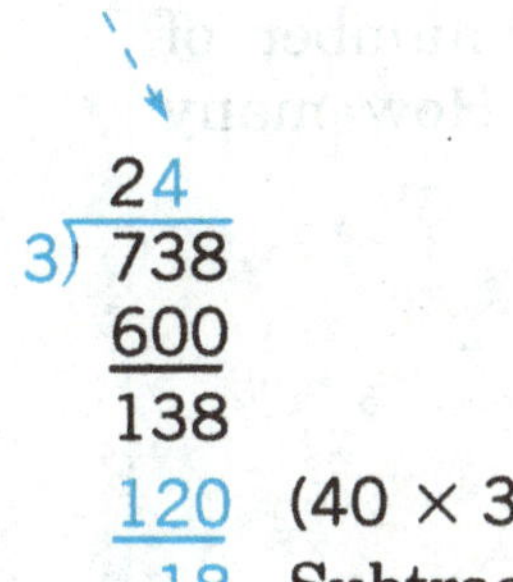

```
    24
3) 738
   600
   138
   120   (40 × 3)
    18   Subtract.
```

×	1	2	3	4	5	6
3	3	6	9	12	15	18

18 ÷ 3 = 6, so the ones digit is 6.

```
                      246
                   3) 738
                      600
                      138
                      120
                       18
                       18   (6 × 3)
remainder (r) --->      0   Subtract.
```

Divide.

	a	*b*	*c*	*d*	*e*
1.	8)96	4)72	6)72	3)81	4)68
2.	2)74	3)87	5)75	7)784	3)768
3.	8)296	9)315	6)252	6)462	5)930

Lesson 2 Problem Solving

Solve each problem.

1. There are 84 scouts in all. Six will be assigned to each tent. How many tents are there?

   There are ________ scouts in all.

   There are ________ scouts in each tent.

   There are ________ tents.

2. Seven people each worked the same number of hours. They worked 91 hours in all. How many hours were worked by each person?

   ________ hours were worked.

   ________ people worked these hours.

   ________ hours were worked by each person.

3. A group of three is a trio. How many trios could be formed with 72 people?

   ________ trios could be formed.

4. A factory shipped 848 cars to 4 cities. Each city received the same number of cars. How many cars were shipped to each city?

   ________ cars were shipped.

   ________ cities received the cars.

   ________ cars were shipped to each city.

5. Malcolm, his brother, and his sister have 702 stamps in all. Suppose each takes the same number of stamps. How many will each get?

   Each will get ________ stamps.

6. There are 6 outs in an inning. How many innings would have to be played to get 348 outs?

   ________ innings would have to be played.

1.	2.
3.	4.
5.	6.

Lesson 3 Division with Remainders

Study how to divide 854 by 4.

×	100	200	300
4	400	800	1200

854

854÷4 is between 200 and 300. The hundreds digit is 2.

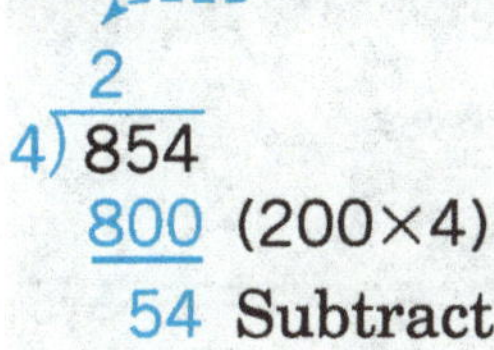

```
    2
4)854
  800  (200×4)
   54  Subtract.
```

×	10	20	30	40
4	40	80	120	160

54

54÷4 is between 10 and 20. The tens digit is 1.

```
   21
4)854
  800
   54
   40   (10×4)
   14   Subtract.
```

×	1	2	3	4	5
4	4	8	12	16	20

14

14÷4 is between 3 and 4. The ones digit is 3.

```
  213 r2
4)854
  800
   54
   40
   14
   12  (3×4)
    2  Subtract.
```


Divide.

	a	*b*	*c*	*d*	*e*
1.	3)82	5)86	4)97	3)76	2)47
2.	7)83	5)69	6)224	4)127	2)380
3.	4)231	5)653	7)962	2)483	6)832

Lesson 3 Problem Solving

Solve each problem.

1. There are 160 packages on 4 large carts. Each cart holds the same number of packages. How many packages are on each cart?

   Each cart has ________ packages.

2. There are 160 packages. To deliver most of the packages, it will take 3 small planes. Each plane will take the same number of packages. How many packages will each plane take? How many packages will be left over?

   Each plane will take ________ packages.

   There will be ________ package(s) left over.

3. Suppose there had been 890 packages to be delivered by 6 planes. Each plane is to take the same number of packages and as many as possible. How many packages will each plane take? How many will be left over?

   Each plane will take ________ packages.

   There will be ________ packages left over.

1.
2.
3.

NAME ____________________

Lesson 4 Checking Division

```
   235
8)1880  <- These should be the same.       Check
  1600                                       235
   280                                        ×8
   240                                      1880
    40
    40
     0
```

```
   178 r2
3) 536  <- These should be the same.       Check
   300                                       178
   236                                        ×3
   210                                       534
    26                                        +2
    24                                       536
     2
```

To check 1880 ÷ 8 = 235,
multiply 235 by 8. The answer should be ________.
To check 536 ÷ 3 = 178 r2,
multiply 178 by 3 and then add 2. The answer should be ________.

CHAPTER 3

Divide. Check each answer.

	a	*b*	*c*
1.	4)1104	8)1760	2)4632
2.	3)379	5)421	4)762
3.	3)1058	6)726	7)2117

Lesson 4 Problem Solving

Solve each problem. Check each answer.

1. How many bags of 7 oranges each can be filled from a shipment of 341 oranges? How many oranges will be left? ________ bags can be filled. ________ oranges will be left.	**1.**
2. Beverly has \$2.38 (238 cents) to buy pencils for 8¢ each. How many pencils can she buy? How many cents will she have left? She can buy ________ pencils. She will have ________ cents left.	**2.**
3. There are 6 stamps in each row. How many complete rows can be filled with 1,950 stamps? How many stamps will be left? ________ rows will be filled. ________ stamps will be left.	**3.**
4. Daphne had 958 pennies. She exchanged them for nickels. How many nickels did she get? How many pennies did she have left? She got ________ nickels. She had ________ pennies left.	**4.**
5. Last year Mr. Gomez worked 1,983 hours. How many 8-hour days was this? How many hours are left? It was ________ 8-hour days. ________ hours are left.	**5.**
6. There are 7,633 points to be divided among Paul, Jeremy, and Paige. Each receives the same number of points. How many points will each receive? How many points will be left? Each student will receive ________ points. ________ point(s) will be left.	**6.**

NAME ____________________

Lesson 5 Estimating Quotients

To estimate a quotient, think of a familiar division fact.

$5\overline{)18}$ $\begin{array}{r} 4 \\ 5\overline{)20} \\ \underline{20} \\ 0 \end{array}$ Subtract	Think of what you can round the dividend (18) to so that it is easy to divide mentally by the divisor (5). The quotient is 4.
$7\overline{)342}$ $\begin{array}{r} 50 \\ 7\overline{)350} \\ \underline{350} \\ 0 \end{array}$ Subtract	Think of what you can round the dividend (342) to so that it is easy to divide mentally by the divisor (7). The quotient is 50.

CHAPTER 3

Estimate each quotient.

	a	*b*	*c*	*d*
1.	$4\overline{)25}$	$8\overline{)46}$	$3\overline{)16}$	$9\overline{)48}$
2.	$3\overline{)175}$	$5\overline{)319}$	$7\overline{)236}$	$6\overline{)474}$
3.	$6\overline{)381}$	$4\overline{)316}$	$8\overline{)652}$	$9\overline{)651}$
4.	$2\overline{)1783}$	$6\overline{)2572}$	$5\overline{)4392}$	$4\overline{)4952}$

Lesson 5 Problem Solving

Solve each problem. Use estimation.

1. Brandon bought cookies to pack in his lunch. He bought a box with 28 cookies. If he packs five cookies in his lunch each day, about how many days will the cookies last?

   The cookies will last for about ________ days.

2. Marta baby-sat for four hours and earned $19. About how much money did Marta earn each hour that she baby-sat?

   Marta earned about ________ each hour that she baby-sat.

3. Bethany rode her bike three days one week. She rode a total of 31 miles. If she rode the same distance each day, about how many miles did she ride each day that week?

   Bethany rode her bike about ________ miles each day that week.

4. Over a four-week period, Terrell earns $354 at his part-time job. About how much money does Terrell earn each week at his part-time job?

   Terrell earns about ________ each week at his part-time job.

5. The school play was performed five times. For the five performances, there was a total of 962 people in attendance. About how many people were in attendance each time the play was performed?

   There were about ________ people in attendance each time the play was performed.

6. An auditorium holds 1,438 people. There are three different sections of seating. Each section has the same number of seats. About how many seats are in each section of the auditorium.

   There are about ________ seats in each section.

1.

2.

3.

4.

5.

6.

NAME ____________________

CHAPTER 3 PRACTICE TEST

Division (2-digit through 4-digit by 1-digit)

Divide.

	a	b	c	d
1.	$4\overline{)96}$	$7\overline{)84}$	$3\overline{)79}$	$5\overline{)68}$
2.	$4\overline{)732}$	$5\overline{)175}$	$7\overline{)615}$	$2\overline{)647}$
3.	$8\overline{)1720}$	$4\overline{)5216}$	$4\overline{)1530}$	$3\overline{)6323}$
4.	$3\overline{)84}$	$6\overline{)76}$	$8\overline{)94}$	$2\overline{)78}$
5.	$4\overline{)1256}$	$3\overline{)6343}$	$5\overline{)1842}$	$6\overline{)7206}$

CHAPTER 3

NAME ______________________________

CHAPTER 4 PRETEST
Division (2-digit through 4-digit by 2-digit)

Divide.

CHAPTER 4

	a	b	c	d
1.	13)78	14)98	12)65	15)95
2.	24)312	37)962	12)586	23)550
3.	27)3564	74)7252	36)2026	34)3830
4.	16)768	52)2724	18)310	14)56
5.	34)4284	53)2120	26)964	11)418

NAME ______________________

Lesson 1 Division (2-digit)

Study how to divide 94 by 13.

Since $10 \times 13 = 130$ and 130 is greater than 94, there is no tens digit.

$13\overline{)94}$

×	1	2	3	4	5	6	7	8
13	13	26	39	52	65	78	91	104

94 is between 91 and 104.
94 ÷ 13 is between 7 and 8.
The *quotient* is 7.

$13\overline{)94}$ quotient 7
91 ← (7×13 = 91)
3 ← (94−91 = 3)

Record the remainder like this.

$13\overline{)94}$ quotient 7 r3
91
3 ← remainder

CHAPTER 4

Divide.

	a	b	c	d	e
1.	$12\overline{)84}$	$13\overline{)78}$	$19\overline{)95}$	$16\overline{)84}$	$14\overline{)98}$
2.	$15\overline{)92}$	$14\overline{)75}$	$16\overline{)74}$	$13\overline{)80}$	$12\overline{)92}$
3.	$17\overline{)68}$	$23\overline{)92}$	$32\overline{)84}$	$18\overline{)72}$	$27\overline{)91}$

Lesson 1 Problem Solving

Solve each problem.

1. The pet store has 84 birds. They have 14 large cages. There are the same number of birds in each cage. How many birds are in each cage?

   ________ birds are in each cage.

2. The pet store also has 63 kittens. There are 12 cages with the same number of kittens in each. The rest of the kittens are in the display window. How many kittens are in each cage? How many kittens are in the display window?

   ________ kittens are in each cage.

   ________ kittens are in the display window.

3. There are 60 guppies in a large tank. If the pet store puts 15 guppies each in a smaller tank, how many smaller tanks will be needed?

   ________ smaller tanks will be needed.

4. There are 72 boxes of pet food on a shelf. The boxes are in rows of 13 each. How many full rows of boxes are there? How many boxes are left over?

   There are ________ full rows of boxes.

   There are ________ boxes left over.

5. There are 80 cages to be cleaned. Each of the store's 19 employees is to clean the same number of cages. The owner will clean any leftover cages. How many cages will each employee clean? How many cages will the owner clean?

   Each employee will clean ________ cages.

   The owner will clean ________ cages.

6. There are 52 puppies. There are 13 cages. If each cage contains the same number of puppies, how many puppies are in each cage?

   There are ________ puppies in each cage.

1.
2.
3.
4.
5.
6.

NAME ______________________

Lesson 2 Division (3-digit)

Study how to divide 219 by 12.

×	10	20	30	40
12	120	240	360	480

219

219 ÷ 12 is between 10 and 20.
The tens digit is 1.

$$\begin{array}{r} 1 \\ 12\overline{)219} \\ 120 \\ \hline 99 \end{array}$$

×	1	2	3	4	5	6	7	8	9
12	12	24	36	48	60	72	84	96	108

99

99 ÷ 12 is between 8 and 9.
The ones digit is 8.

$$\begin{array}{r} 18\text{ r}3 \\ 12\overline{)219} \\ 120 \\ \hline 99 \\ 96 \\ \hline 3 \end{array}$$

Divide.

	a	*b*	*c*	*d*	*e*
1.	$13\overline{)351}$	$16\overline{)256}$	$17\overline{)323}$	$14\overline{)490}$	$12\overline{)814}$
2.	$26\overline{)316}$	$31\overline{)413}$	$17\overline{)212}$	$24\overline{)360}$	$28\overline{)564}$

CHAPTER 4

Lesson 2 Problem Solving

Solve each problem.

1. There are 448 reams of paper in the supply room. Fourteen reams are used each day. At that rate, how many days will the supply of paper last?

   The supply of paper will last ________ days.

2. There are 338 cases on a truck. The truck will make 12 stops and leave the same number of cases at each stop. How many cases will be left at each stop? How many cases will still be on the truck?

   ________ cases will be left at each stop.

   ________ cases will still be on the truck.

3. There are 582 tickets to be sold. Each of 24 students is to receive the same number of tickets and as many as possible. The teacher is to sell any tickets left over. How many tickets is each student to sell? How many is the teacher to sell?

   Each student is to sell ________ tickets.

   The teacher is to sell ________ tickets.

4. A machine operated 38 hours and produced 988 parts. The same number of parts was produced each hour. How many parts were produced each hour?

   ________ parts were produced each hour.

5. After 24 hours, the machine in problem **4** had produced 576 parts. The same number of parts was produced each hour. How many parts did the machine produce each hour?

   ________ parts were produced each hour.

6. Suppose the machine in problem **4** was operated 19 hours. During this time 988 parts were produced. The same number of parts was produced each hour. How many were produced each hour?

   ________ parts were produced each hour.

1.
2.
3.
4.
5.
6.

NAME ________________

Lesson 3 Division (3-digit)

$$\begin{array}{r} 8 \text{ r2} \\ 12\overline{)98} \\ \underline{96} \\ 2 \end{array}$$

These should be the same.

Check

$$\begin{array}{r} 8 \\ \underline{\times 12} \\ 16 \\ \underline{80} \\ 96 \\ \underline{+2} \\ 98 \end{array}$$

To check 98 ÷ 12 = 8 r2, multiply 8 by ______ and add ______ to that product.

The answer should be ______.

$$\begin{array}{r} 12 \\ 34\overline{)408} \\ \underline{340} \\ 68 \\ \underline{68} \\ 0 \end{array}$$

These should be the same.

Check

$$\begin{array}{r} 12 \\ \underline{\times 34} \\ 48 \\ \underline{360} \\ 408 \end{array}$$

To check 408 ÷ 34 = 12, multiply 12 by ______. The answer should be ______.

CHAPTER 4

Divide. Check each answer.

	a	*b*	*c*
1.	$16\overline{)88}$	$14\overline{)84}$	$23\overline{)94}$
2.	$19\overline{)114}$	$36\overline{)756}$	$32\overline{)836}$
3.	$25\overline{)330}$	$36\overline{)672}$	$45\overline{)810}$

Lesson 3 Problem Solving

Solve each problem. Check each answer.

1. Lucinda had 59 cents to buy pencils that cost 14 cents each. How many pencils could she buy? How many cents would she have left?

   She could buy ________ pencils.

   She would have ________ cents left.

2. The grocer has 98 cans of beans to put on a shelf. He thinks he can put 16 cans in each row. If he does, how many rows will he have? How many cans will be left?

   He will have ________ rows.

   ________ cans will be left.

3. The grocer in problem **2** could only put 13 cans in each row. How many rows does he have? How many cans are left?

   He has ________ rows.

   ________ cans are left.

4. There are 774 cartons ready for shipment. Only 27 cartons can be shipped on each truck. How many full truckloads will there be? How many cartons will be left?

   There will be ________ full loads.

   ________ cartons will be left.

5. There are 605 books in the storage room. There are the same number of books in each of 17 full boxes and the rest in an extra box. How many books are in each full box? How many books are in the extra box?

   ________ books are in each full box.

   ________ books are in the extra box.

1.

2.

3.

4.

5.

NAME ____________________

Lesson 4 Division (4-digit)

Study how to divide 8550 by 25.

×	100	200	300	400
25	2500	5000	7500	10000

8550

The hundreds digit is 3.

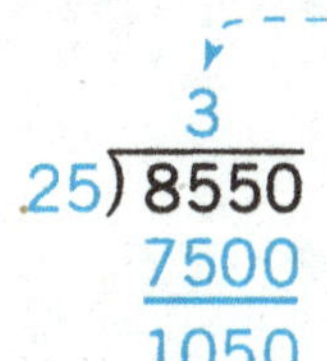

×	10	20	30	40	50
25	250	500	750	1000	1250

1050

The tens digit is 4.

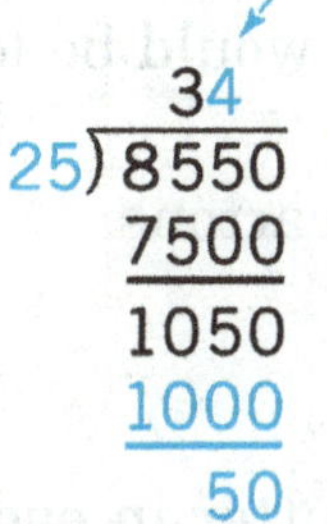

×	1	2
25	25	50

50

The ones digit is 2.

$$\begin{array}{r} 342 \\ 25\overline{)8550} \\ \underline{7500} \\ 1050 \\ \underline{1000} \\ 50 \\ \underline{50} \\ 0 \end{array}$$

CHAPTER 4

Divide.

	a	*b*	*c*	*d*
1.	$32\overline{)5280}$	$43\overline{)6751}$	$26\overline{)6318}$	$75\overline{)9150}$
2.	$42\overline{)8956}$	$31\overline{)9875}$	$23\overline{)3844}$	$63\overline{)9008}$
3.	$35\overline{)1960}$	$75\overline{)3900}$	$63\overline{)2656}$	$27\overline{)1430}$

Lesson 4 Problem Solving

Solve each problem.

1. A truck is loaded with 8,073 kilograms of food. Each case of food weighs 23 kilograms. How many cases are on the truck?

   _________ cases are on the truck.

2. During an 8-hour shift, one machine was able to package 8,215 boxes of rice. These boxes were packed 24 to a carton. How many full cartons of rice would this be? How many boxes would be left?

   There would be _________ full cartons.

   _________ boxes would be left.

3. The bakery uses 75 pounds of butter in each batch of butter-bread dough. How many batches of dough could be made with 6,300 pounds of butter?

   _________ batches of dough could be made.

4. There are 2,030 students in school. How many classes of 28 students each could there be? How many students would be left?

   There could be _________ full classes.

   _________ students would be left.

5. In 27 days 3,888 gallons of oil were used. The same amount of oil was used each day. How much oil was used each day?

   _________ gallons were used each day.

6. There are 5,100 parts to be packed. The parts are to be packed 24 to a box. How many boxes can be filled? How many parts will be left?

   _________ full boxes will be packed.

   _________ parts will be left.

1.
2.
3.
4.
5.
6.

Lesson 5 Division (2-, 3-, and 4-digit)

Divide.

	a	*b*	*c*	*d*
1.	$28\overline{)776}$	$42\overline{)5176}$	$19\overline{)95}$	$33\overline{)133}$
2.	$12\overline{)2606}$	$22\overline{)6754}$	$24\overline{)792}$	$11\overline{)1716}$
3.	$14\overline{)84}$	$89\overline{)801}$	$75\overline{)753}$	$16\overline{)2616}$
4.	$75\overline{)6375}$	$23\overline{)5543}$	$25\overline{)8000}$	$25\overline{)800}$
5.	$15\overline{)6009}$	$60\overline{)1860}$	$20\overline{)7020}$	$48\overline{)1704}$

CHAPTER 4

Lesson 5 Problem Solving

ORDER FORM
6,912 Zanappas

Solve each problem.

1. An order was received for 6,912 zanappas. Machine A can produce the zanappas in 12 hours. At that rate, how many zanappas would be produced each hour?

   ________ zanappas would be produced each hour.

2. It would take Machine B 24 hours to produce the zanappas needed to fill the order. At that rate, how many zanappas would be produced each hour?

   ________ zanappas would be produced each hour.

3. Machine C could produce the zanappas needed to fill the order in 48 hours. At that rate, how many zanappas could be produced each hour?

   ________ zanappas could be produced each hour.

4. How many zanappas could be produced if all three machines operated for a period of 8 hours?

   ________ zanappas could be produced.

1.	2.
3.	4.

NAME ______________________

CHAPTER 4 PRACTICE TEST

Division (2-digit through 4-digit by 2-digit)

Divide.

	a	b	c	d
1.	12)72	13)89	11)94	17)68
2.	17)265	11)858	31)961	12)506
3.	36)4366	42)1890	73)3934	14)2184
4.	13)169	26)3175	16)75	36)144
5.	54)1458	25)2095	28)573	42)99

CHAPTER 4

NAME ______________________

CHAPTER 5 PRETEST
Division (4- and 5-digit by 2-digit)

Divide.

	a	*b*	*c*	*d*
1.	25)7 5	25)7 5 0	25)7 5 0 0	25)7 5 0 0 0
2.	38)4 2 5 6	17)4 0 3 3	33)7 3 2 6	25)2 1 4 5
3.	42)8 9 5 2 3	16)9 7 9 7 8	25)6 2 9 4 0	15)3 1 7 6 2
4.	27)1 2 2 0 4	48)2 7 6 4 8	62)1 9 6 6 4	72)3 1 9 6 8

CHAPTER 5

Lesson 1 Division (5-digit)

Study how to divide 24,567 by 12.

×	1000	2000	3000
12	12000	24000	36000

24567

The thousands digit is 2.

```
      2
12) 24567
    24000
      567
```

×	100	200
12	1200	2400

567 ÷ 12 is less than 100. The hundreds digit is 0.

```
      20
12) 24567
    24000
      567
```

×	30	40	50
12	360	480	600

567

The tens digit is 4.

```
      204
12) 24567
    24000
      567
      480
       87
```

×	6	7	8
12	72	84	96

87

The ones digit is 7.

```
      2047 r3
12) 24567
    24000
      567
      480
       87
       84
        3
```

Divide.

	a	b	c	d
1.	36)4500	26)8430	92)7911	25)3575
2.	24)77184	92)39754	56)69104	23)17342

CHAPTER 5

Lesson 1 Problem Solving

Solve each problem.

1. In 27 days, 6,939 orders were filled. The same number of orders was filled each day. How many orders were filled each day?

   ________ orders were filled each day.

2. Yesterday 5,650 school children came in buses to visit the museum. How many full bus loads of students were there if 75 students make up a full load? How many students were on the partially filled bus?

   There were ________ full bus loads.

   ________ students were on the partially filled bus.

3. The inventory slip shows that there are 7,840 pairs of stockings in the warehouse. There are 32 pairs in each box. How many boxes of stockings should there be in the warehouse?

   There should be ________ boxes of stockings.

4. A factory produced 7,605 zimbits yesterday. The zimbits are packed 24 to a box. How many full boxes of zimbits were produced? How many zimbits were left?

   There were ________ full boxes.

   ________ zimbits were left.

5. The grandstand is separated into 16 sections. Each section has the same number of seats. There are 8,640 seats in all. How many seats are in each section?

   There are ________ seats in each section.

6. Suppose there were 9,600 seats in the grandstand in problem **5**. How many seats would be in each section?

   There would be ________ seats in each section.

1.
2.
3.
4.
5.
6.

NAME ____________________

Lesson 2 Division (5-digit)

Study how to divide 24,205 by 75.

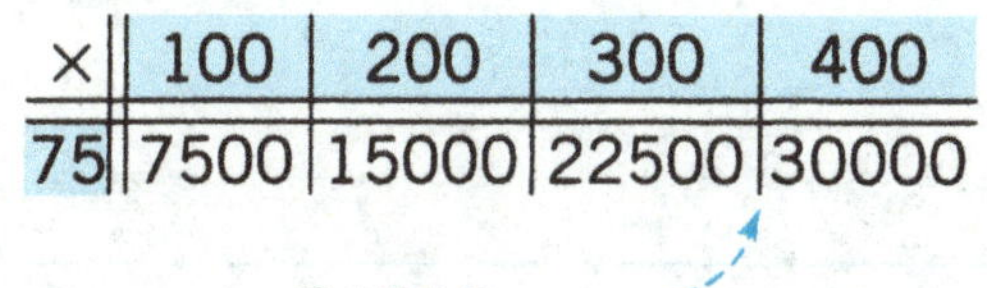

×	100	200	300	400
75	7500	15000	22500	30000

24205

The hundreds digit is 3.

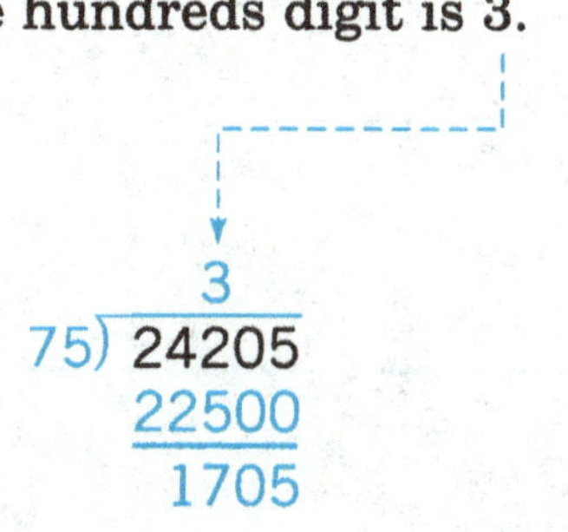

×	10	20	30	40
75	750	1500	2250	3000

1705

The tens digit is 2.

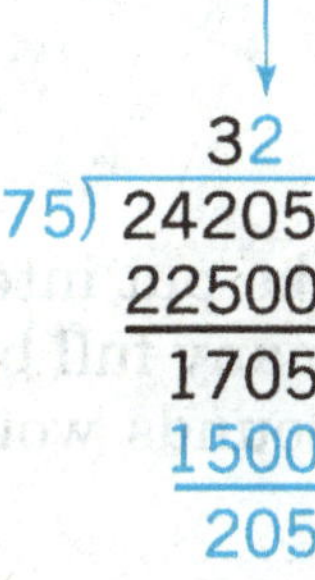

×	1	2	3	4
75	75	150	225	300

205

The ones digit is 2.

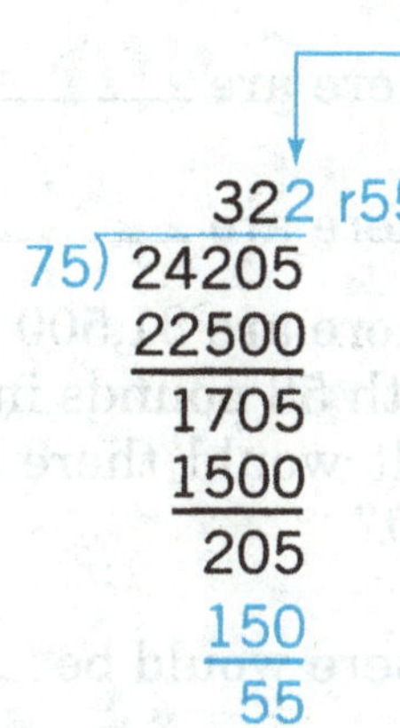

CHAPTER 5

Divide.

	a	*b*	*c*	*d*
1.	43) 1 7 7 1 6	64) 3 2 7 6 8	27) 2 2 0 0 5	28) 6 0 0 8 8
2.	33) 2 7 3 1 3	31) 9 6 8 4 3	43) 8 9 8 0 0	59) 4 1 6 4 5

Lesson 2 Problem Solving

Solve each problem.

1. A bus can carry 86 passengers. How many such buses would be needed to carry 20,898 passengers?

   ________ buses would be needed.

2. There are 16 ounces in one pound. How many pounds are there in 39,238 ounces? How many ounces are left?

   There are ________ pounds.

   There are ________ ounces left.

3. There are 31,500 pounds of salt to be put into bags with 58 pounds in each bag. How many full bags of salt would there be? How many pounds would be left?

   There would be ________ full bags.

   ________ pounds would be left.

4. It takes 72 hours for one machine to produce 14,616 parts. The machine produces the same number of parts each hour. How many parts does it produce each hour?

   It produces ________ parts each hour.

5. Suppose the machine in problem **4** could produce the parts in 36 hours. How many parts would it produce each hour?

   It would produce ________ parts each hour.

6. Suppose the machine in problem **4** could produce the parts in 18 hours. How many parts would it produce each hour?

   It would produce ________ parts each hour.

7. Suppose the machine in problem **4** could produce the parts in 12 hours. How many parts would it produce each hour?

   It would produce ________ parts each hour.

1.
2.
3.
4.
5.
6.
7.

NAME ____________________

Lesson 3 Checking Division

```
    2543 r8
16)40696
   32000
    8696
    8000
     696
     640
      56
      48
       8
```

These should be the same.

Check

```
  2543
  ×16
 15258
 25430
 40688
    +8
 40696
```

To check 40696 ÷ 16 = 2543 r8, multiply 2543 by ________ and then add _____ to this product. The answer should be __________ .

CHAPTER 5

Divide. Check each answer.

	a	*b*
1.	47)99932	54)33100
2.	38)27590	46)38277
3.	75)95100	24)30900

Lesson 3 Problem Solving

Solve each problem. Check each answer.

1. There are 35 gates into the stadium and 15,330 people attended the game. The same number entered through each gate. How many entered through each gate?

   _________ people entered through each gate.

2. A garage used 16,434 liters of oil in 83 days. The same amount of oil was used each day. How much oil was used each day?

   _________ liters were used each day.

3. During 6 months, 77 employees worked 67,639 hours. Suppose each employee worked the same number of hours. How many hours did each work? How many hours would be left?

   Each employee worked _________ hours.

   _________ hours are left.

4. Ninety-five containers of the same size were filled with a total of 82,840 kilograms of coal. How many kilograms of coal were in each container?

   _________ kilograms were in each container.

5. There are 46,963 students attending 52 schools in the city. Suppose the same number attend each school. How many students would attend each school? How many would be left?

   _________ students would attend each school.

   _________ students would be left.

6. Suppose there were twice as many students in problem **5.** How many students would attend each school? How many would be left?

   _________ students would attend each school.

   _________ students would be left.

1.
2.
3.
4.
5.
6.

Lesson 4 Division (4- and 5-digit)

Divide.

	a	*b*	*c*	*d*
1.	38)72	23)601	32)4640	34)43877
2.	24)54	24)540	24)5400	24)54000
3.	12)87	21)168	42)1491	38)21584
4.	87)95	24)369	75)6005	45)30605

Lesson 4 Problem Solving

Solve each problem.

1. Hannah is to read 228 pages in 4 sessions. She will read the same number of pages each session. How many pages will she read each session?

   She will read ________ pages each session.

   1.

2. The square of a number is found by multiplying the number by itself. Matthew said that 2,916 is the square of 54. Is he right?

   Matthew ________ right.

   2.

3. The astronauts are now 8,640 minutes into their flight. How many hours would this be? How many days?

   It would be ________ hours.

   It would be ________ days.

   3.

4. In five hours 15,190 cans came off the assembly line. There are 88 cans packed in each carton. How many full cartons are there? How many cans are in the partially filled carton?

   There are ________ full cartons.

   There are ________ cans in the partial carton.

   4.

5. A satellite has just completed its 94th orbit. It has been in orbit for 13,160 hours. How long does it take to make a complete orbit?

   It takes ________ hours to make one orbit.

   5.

6. How long will the satellite in problem **5** have been in orbit after it has completed its 100th orbit?

   It will have been in orbit ________ hours.

   6.

NAME ______________________

CHAPTER 5 PRACTICE TEST

Division (4- and 5-digit by 2-digit)

CHAPTER 5

Divide.

	a	b	c	d
1.	97)873	56)952	70)2870	63)6615
2.	31)8308	41)5043	11)1232	77)9831
3.	32)23744	93)31657	51)21483	43)31605
4.	25)23375	17)34096	37)65510	77)92324
5.	35)35035	25)10025	31)93006	13)10413

NAME ____________

CHAPTER 6 PRETEST
Money

In each dollar amount, ring the digit in the given place value.

	a	*b*
1.	\$345.06; hundredths place	\$705.98; tenths place
2.	\$4,587.91; tens place	\$135.74; hundreds place

Express each dollar amount in standard form.

3. eighty-three dollars and seventy-five cents ____________

4. seven hundred sixty-three dollars and forty-nine cents ____________

CHAPTER 6

Add.

	a	*b*	*c*	*d*	*e*
5.	\$2.94 +6.18	\$32.05 +18.76	\$267.31 +46.98	\$319.46 +229.35	\$4864.11 +2417.39

Subtract.

6.	\$9.41 −3.18	\$33.29 −28.47	\$705.92 −56.35	\$468.13 −125.26	\$5671.48 −3489.17

Multiply.

7.	\$1.38 × 4	\$8.63 × 7	\$37.81 × 5	\$5.49 × 21	\$55.19 × 38

Divide.

8.	3)\$0.87	8)\$9.84	6)\$34.56	14)\$7.56	47)\$99.64

NAME ______________________

Lesson 1 Place Value

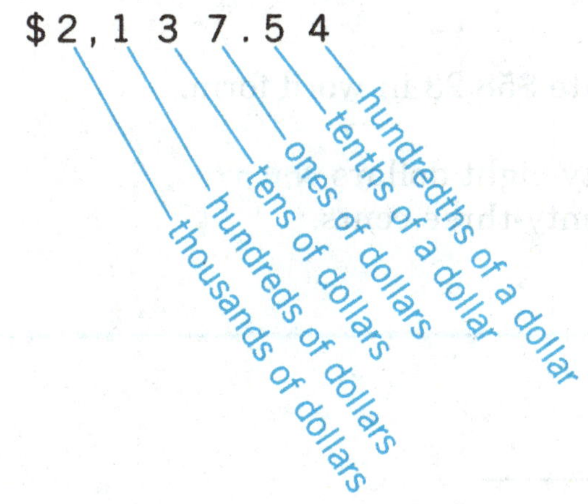

In $6,043.25, what number is in the tenths place?

2 is in the tenths place.

CHAPTER 6

In each dollar amount, ring the digit in the given place value.

	a	*b*
1.	$194.28; ones place	$3,829.76; hundreds place
2.	$3,581.06; hundredths place	$815.49; tens place
3.	$264.97; hundreds place	$5,216.38; tenths place
4.	$6,804.15; tenths place	$197.32; ones place
5.	$3,248.05; thousands place	$8,213.45; hundredths place
6.	$5,810.64; hundredths place	$2,183.67; thousands place
7.	$315.42; ones place	$467.03; tenths place
8.	$7,241.36; tens place	$3,425.10; ones place
9.	$5,316.24; hundreds place	$365.42; hundreds place
10.	$516.37; tenths place	$9,216.35; hundredths place
11.	$4,256.38; hundredths place	$8,093.17; tens place
12.	$1,834.90; thousands place	$6,314.25; thousands place

NAME ____________________

Lesson 2 Writing Money

Write "two hundred eighteen dollars and thirty-seven cents" in standard form.

$218.37

Write $58.23 in word form.

Fifty-eight dollars and twenty-three cents.

Express each dollar amount in standard form.

1. thirty-seven dollars and twenty-six cents ________
2. seventy-two dollars and sixty-one cents ________
3. four hundred twenty-one dollars and thirty-five cents ________
4. five hundred forty-four dollars and thirteen cents ________
5. nine hundred eighty-one dollars and ninety cents ________
6. two thousand, seven hundred thirty-six dollars and forty-five cents ________

Express each dollar amount in word form.

7. $41.57 ____________________
8. $30.24 ____________________
9. $652.74 ____________________
10. $426.13 ____________________
11. $703.89 ____________________
12. $3,950.21 ____________________

NAME ______________________

Lesson 3 Addition of Money

To add money, line up the decimal points. Then add from right to left.

$1.29	$1.29 (carry 1)	$1.29 (carry 1, 1)	$1.29
+0.84	+0.84	+0.84	+0.84
	3	.13	$2.13

Add.

	a	b	c	d	e
1.	$0.42 +1.83	$6.90 +2.63	$3.45 +7.32	$6.08 +0.73	$3.12 +9.76
2.	$21.54 +18.13	$42.81 +37.52	$15.64 +12.21	$46.81 +19.84	$62.18 +73.48
3.	$623.65 +82.79	$483.21 +21.19	$265.95 +38.41	$80.21 +363.14	$28.35 +917.64
4.	$648.15 +794.05	$345.61 +725.88	$483.85 +297.65	$724.31 +613.55	$511.29 +180.46
5.	$2316.25 +6108.34	$4866.23 +1585.50	$4786.03 +4671.29	$5024.87 +1094.71	$1358.01 +8274.81
6.	$52.64 77.64 +29.05	$13.29 264.08 +38.27	$243.25 71.78 +883.40	$2268.06 411.52 +4186.54	$4545.19 6133.41 +8009.77

CHAPTER 6

Lesson 3 Problem Solving

Solve each problem.

1. At the convenience store, Danielle bought a gallon of milk that cost $2.85 and a bag of pretzels that cost $1.99. How much money did Danielle spend at the convenience store?

   Danielle spent ________ at the convenience store.

2. On Friday night, Wynona earned $11.75 baby-sitting. On Saturday afternoon, she earned $5.50 helping her neighbor clean her house. How much money did Wynona earn on Friday and Saturday?

   Wynona earned ________________ on Friday and Saturday.

3. Justin bought a bike that cost $128.50 and a helmet that cost $29.75. How much money did Justin spend on the bike and helmet?

   Justin spent ________ on the bike and helmet.

4. Jamar went to the deli for lunch. He bought a sandwich for $2.75, a bag of chips for $0.80, and a fruit juice for $1.35. How much money did Jamar spend at the deli for lunch?

   Jamar spent ________ at the deli for lunch.

5. Kyle has money in a checking account and a savings account. He has $348.12 in his checking account and $1,069.57 in his savings account. How much money does Kyle have in his two accounts?

   Kyle has __________ in his two accounts.

6. Ramona is a salesperson. In January, her sales total was $6,458.24. In February, her sales total was $8,058.75. What was Ramona's sales total for January and February?

   Ramona's sales total for January and February was __________.

1.

2.

3.

4.

5.

6.

NAME ______

Lesson 4 Subtraction of Money

To subtract money, line up the decimal points. Then subtract from right to left.

\$15.65 −7.49	5 15 \$15.65 −7.49 6	5 15 \$15.65 −7.49 .16	0 15 5 15 \$15.65 −7.49 \$8.16

Subtract.

	a	b	c	d	e
1.	\$6.84 −1.19	\$9.59 −5.75	\$5.02 −3.18	\$4.53 −3.26	\$7.27 −6.98
2.	\$45.34 −17.50	\$12.80 −7.41	\$78.46 −34.85	\$41.63 −39.45	\$56.42 −34.13
3.	\$345.91 −97.37	\$469.25 −57.36	\$427.63 −90.47	\$727.44 −78.34	\$410.65 −99.27
4.	\$764.33 −551.92	\$604.15 −538.22	\$926.40 −674.12	\$541.73 −336.95	\$831.92 −624.51
5.	\$3068.15 −715.94	\$5943.05 −843.16	\$3128.65 −461.08	\$4216.38 −530.08	\$8003.69 −742.31
6.	\$3450.18 −2290.46	\$5024.88 −2656.28	\$8124.77 −7653.25	\$1192.06 −1086.33	\$9624.08 −3782.14

CHAPTER 6

Lesson 4 Problem Solving

Solve each problem.

1. On Wednesday, Caroline spent $3.80 for lunch. On Thursday, she spent $5.25 for lunch. How much more money did Caroline spend on Thursday for lunch than on Wednesday?

   Caroline spent ______ more on Thursday for lunch than on Wednesday.

2. At Pete's Pizzeria, a large pizza costs $12.50. A medium pizza costs $8.75. How much more does a large pizza cost than a medium pizza at Pete's Pizzeria?

   A large pizza costs ______ more than a medium pizza at Pete's Pizzeria.

3. Deontay and Patrick went to a music store. Deontay bought a CD that cost $17.50. Patrick bought a CD that cost $12.99. How much more did Deontay spend than Patrick?

   Deontay spent ______ more for his CD than Patrick.

4. Keung earns $13.50 for mowing his grandmother's lawn. He earns $15.00 for mowing his neighbor's lawn. How much more money does Keung earn for mowing his neighbor's lawn than he does for mowing his grandmother's lawn?

   Keung earns ______ more for mowing his neighbor's lawn.

5. Rachel has $564.12 in her savings account. Mitchell has $392.46 in his savings account. How much more money does Rachel have in her savings account than Mitchell?

   Rachel has ______ more in her savings account.

6. Jasmine bought a new car for $9,867.50. She sold her old car for $3,575.00. She used all the money from selling her old car as a deposit for her new car. After the deposit, how much more money did Jasmine have to pay for her new car?

   Jasmine had to pay ________ more for her new car.

1.

2.

3.

4.

5.

6.

NAME ____________________

Lesson 5 Multiplication of Money

Multiply money the same way you multiply with whole numbers.

```
  3          5 3          5 3
$2.85      $2.85        $2.85 ← 2 decimal places
  ×6         × 6          × 6
     0         10      $17.10 ← 2 decimal places
```

Be sure to include the dollar sign and decimal point in your answer.

Multiply 835 by 4. Then multiply 835 by 20. Add. Then write the dollar sign and decimal point.

```
2 decimal places →  $8.35
                     ×24
                    3340 }
                   16700 } Add.
2 decimal places → $200.40
```

Multiply.

	a	b	c	d	e
1.	$1.15 × 3	$3.68 × 8	$6.52 × 7	$1.33 × 9	$9.11 × 5
2.	$25.48 × 7	$74.12 × 2	$65.43 × 5	$45.35 × 4	$39.04 × 9
3.	$6.27 × 18	$9.25 × 34	$8.45 × 52	$31.60 × 94	$79.25 × 21
4.	$65.80 × 35	$33.64 × 68	$27.81 × 41	$13.62 × 185	$84.67 × 315

CHAPTER 6

Lesson 5 Problem Solving

Solve each problem.

1. Apples are on sale for $0.99 per pound. Zach bought 3 pounds of apples. How much money did Zach spend on the apples?

   Zach spent ________ on the apples.

2. At the concession stand, hot dogs cost $1.25. Mr. Garcia bought 6 hot dogs for his family. How much did Mr. Garcia pay for the hot dogs?

   Mr. Garcia paid ________ for the hot dogs.

3. Natasha earns $4.25 an hour for baby-sitting. If Natasha baby-sits for 5 hours, how much money does she earn?

   Natasha earns ________ for baby-sitting 5 hours.

4. Michael bought 15 packs of baseball cards. If each pack costs $1.95, how much money did Michael spend on the baseball cards?

   Michael spent ________ on the baseball cards.

5. Jenny earns $325.72 each month at her part-time job. How much money does Jenny earn in one year at her job?

   Jenny earns ____________ in one year.

6. At a rental car company, it costs $26.80 per day to rent a mid-size car. Lamar rented a mid-size car from this company for 14 days. How much did Lamar spend for the rental car?

   Lamar spent ________ for the rental car.

7. A company had a summer picnic for its employees. The food for the picnic was catered. The food cost $3.75 per person. If there were 264 people at the picnic, what was the total cost for the food?

   The total cost for the food at the company picnic was ____________.

1.
2.
3.
4.
5.
6.
7.

NAME ______________________

Lesson 6 Division of Money

When dividing money, place the decimal point in the quotient over the decimal point in the dividend. Then divide as if you were dividing whole numbers.

6)$2.58	$0.4 6)$2.58 240	$0.4 6)$2.58 240 18 (Subtract.)	$0.43 6)$2.58 240 18 (Subtract.) 18 0 (Subtract.)

CHAPTER 6

Divide.

	a	*b*	*c*	*d*	*e*
1.	4)$0.68	8)$1.04	6)$1.32	5)$0.90	7)$3.01
2.	8)$4.24	3)$2.88	7)$6.09	4)$5.32	9)$7.38
3.	7)$20.86	5)$33.85	4)$73.72	3)$74.97	6)$75.48
4.	23)$8.28	16)$7.20	34)$8.84	12)$5.28	28)$9.52
5.	42)$90.72	85)$55.25	33)$98.01	62)$96.10	31)$85.87

Lesson 6 Problem Solving

Solve each problem.

1. Kathy bought 4 cucumbers at the produce store for $2.36. How much did each cucumber cost?

   Each cucumber cost ______ at the produce store.

2. Nestor earned $14.00 for helping his elderly neighbor do yard work for 4 hours. How much did Nestor earn each hour?

   Nestor earned ______ each hour he helped his neighbor.

3. Maria spent $14.75 on five packs of thank-you cards. How much did each pack of thank-you cards cost?

   Each pack of thank-you cards cost ______.

4. For a school fund-raiser, Jeremy sold 34 candy bars. He collected $42.50 for the candy bars he sold. How much did each candy bar cost?

   Each candy bar cost ______.

5. Tina bought six pairs of socks for $14.94. How much did each pair of socks cost?

   Each pair of socks cost ______.

6. Naoko bought three new pairs of khaki pants for work. He spent $71.85 for all three pairs. How much did each pair of pants cost?

   Each pair of pants cost ______.

7. The community service club had a fund-raiser to raise money for three local charities. They raised $98.61. If they split the money equally among the three charities, how much money will each charity receive?

   Each charity will receive ______ from the community service club.

1.
2.
3.
4.
5.
6.
7.

NAME ____________________

Lesson 7 Problem Solving

Mariana bought a pair of shoes for $26.85. Two weeks later, Mariana's friend Lauren bought the same pair of shoes for $21.58. How much more did Mariana pay for the pair of shoes than Lauren?

Are you to add or subtract? subtract

Mariana paid $5.27 more for the shoes than Lauren.

Subtract the two amounts to find out how much more Mariana paid for the shoes than Lauren.

7 15 $26.85 −21.58 7	7 15 $26.85 −21.58 .27	7 15 $26.85 −21.58 $5.27

Answer each question.

1. The soccer coach bought five soccer balls for his team. If each soccer ball costs $12.95, how much will the coach spend for five soccer balls?

   Are you to multiply or divide? ________

   How much will the coach spend for five soccer balls? ________

1.

2. Madeline bought a picture frame that cost $7.65. She paid with a $10 bill. How much change did Madeline get back?

   Are you to add or subtract? ________

   How much change did Madeline get back? ________

2.

3. Charlie bought six tickets to the school play for $22.50. How much did each ticket cost?

   Are you to multiply or divide? ________

   How much did each ticket cost? ________

3.

4. Janessa went to the pet store and bought a new collar for her dog that cost $5.95. She also bought a bag of food that cost $12.16. How much money did Janessa spend at the pet store?

   Are you to add or subtract? ________

   How much money did Janessa spend at the pet store? ________

4.

CHAPTER 6

Lesson 7 Problem Solving

Answer each question.

1. Juanita earns $6.22 per hour at her part-time job. Nicole earns $5.17 per hour at her part-time job. How much more per hour does Juanita earn at her part-time job than Nicole?

   Are you to add or subtract? ____________

   How much more per hour does Juanita earn at her part-time job than Nicole? ________

   1.

2. Brian spent $14.37 for three pounds of ham at the deli. How much does one pound of ham cost at the deli?

   Are you to multiply or divide? ________

   How much does one pound of ham cost at the deli? ________

   2.

3. Clara went shopping at the mall. She bought a sweater that cost $24.99, a pair of sunglasses that cost $12.65, and a pair of earrings that cost $6.20. How much money did Clara spend at the mall?

   Are you to add or subtract? ______

   How much did Clara spend at the mall? ______

   3.

4. At the school cafeteria, a slice of pizza costs $1.35 and a hamburger costs $2.60. How much more does a hamburger cost than a slice of pizza at the school cafeteria?

   Are you to add or subtract? __________

   How much more does a hamburger cost than a slice of pizza at the school cafeteria? ______

   4.

5. Rashad bought an 8-pound turkey at the grocery store. It was on sale for $1.79 per pound. How much did Rashad spend on the turkey?

   Are you to multiply or divide? __________

   How much did Rashad spend on the turkey? ______

   5.

NAME ____________

CHAPTER 6 PRACTICE TEST
Money

In each dollar amount, ring the digit in the given place value.

	a	*b*
1.	$903.48; ones place	$673.19; hundredths place
2.	$8,364.82; tenths place	$1,813.06; thousands place

Express each dollar amount in standard form.

3. thirty-four dollars and seventy-two cents ____________

4. two hundred fifteen dollars and ninety-one cents ____________

Add.

	a	*b*	*c*	*d*	*e*
5.	$2.59 +8.43	$27.64 +71.09	$684.05 +83.47	$187.46 +456.45	$1845.54 +9453.72

Subtract.

6.	$7.94 −4.57	$79.18 −53.22	$619.06 −74.25	$914.47 −294.51	$7064.38 −2271.45

Multiply.

7.	$1.07 × 6	$6.19 × 4	$29.67 × 8	$7.85 × 36	$24.92 × 28

Divide.

8.	6)$0.96	7)$9.94	5)$56.85	26)$9.10	21)$90.93

CHAPTER 6

NAME ______________________

CHAPTER 7 PRETEST
Graphs and Averages

Use the bar graph to answer each question.

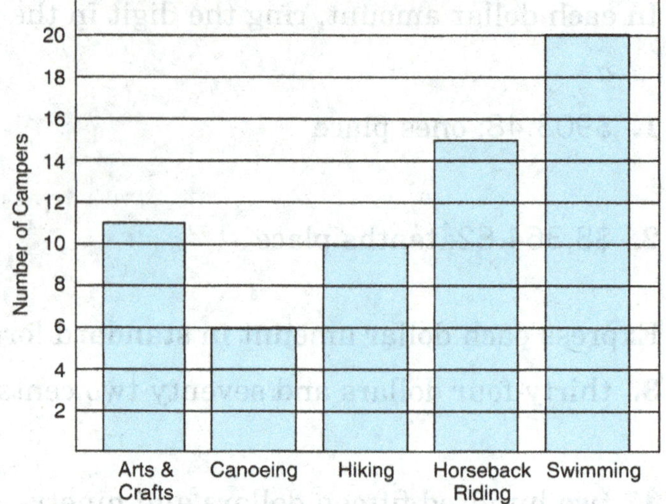

1. How many campers chose canoeing as their favorite activity? ________
2. How many campers chose horseback riding as their favorite activity? ________
3. How many more campers chose arts and crafts as their favorite activity than hiking? ________
4. How many fewer campers chose canoeing as their favorite activity than swimming? ________

Use the line graph to answer each question.

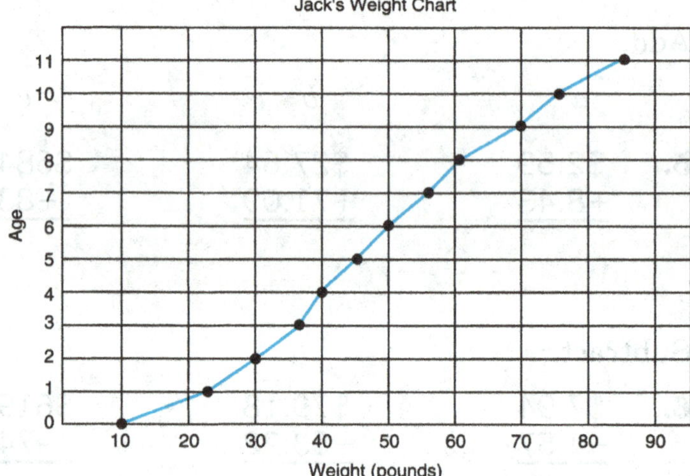

5. How much did Jack weigh when he was 2 years old? ______________
6. How much did Jack weigh when he was 10 years old? ______________
7. How much more did Jack weigh when he was 9 years old than when he was 5 years old? ______________

Find the mean, median, mode, and range of each set of numbers.

	a		*b*	
8. 3, 7, 3, 8, 9	mean: ________	13, 19, 15, 22, 14, 19, 17	mean: ________
	median: ________		median: ________
	mode: ________		mode: ________
	range: ________		range: ________

9. A bag contains nine marbles. Four marbles are blue. Two marbles are green. Two marbles are red. One marble is orange. What is the probability of selecting an orange marble without looking? ________

NAME ______________________

Lesson 1 Bar Graphs

The **bar graph** shows the number of scooters that were sold at a particular store from January to June of one year.

How many scooters were sold in April?

Locate the top edge of the bar that represents April. Follow this to the left to locate its value on the vertical bar.

In April, ___25___ scooters were sold.

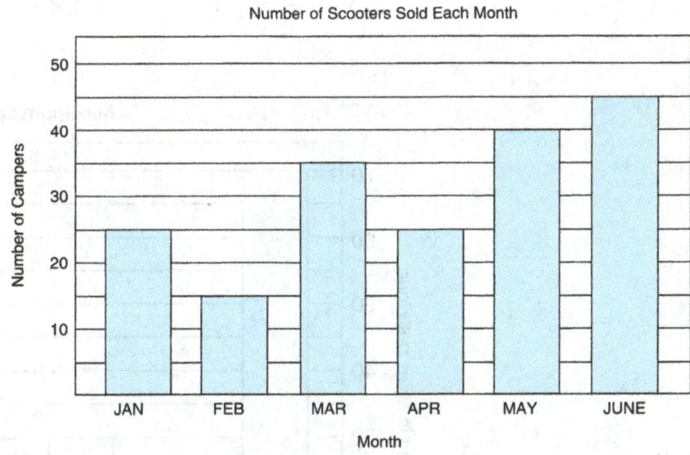

CHAPTER 7

Use the bar graph to answer each question.

1. How many scooters were sold in January? ________

2. How many scooters were sold in March? ________

3. How many scooters were sold in June? ________

4. In which two months were the same number of scooters sold?

 ____

5. How many more scooters were sold in May than in April? ________

6. How many fewer scooters were sold in February than in January? ________

7. What is the minimum number of scooters that need to be sold in July to have more sales than in June? ________

8. Between which two consecutive months did the sales increase the most?

 ____

9. Between which two consecutive months did the sales decrease the most?

 ____

10. The sales goal for July through December is to sell more scooters than were sold in January through June. What is the minimum number of scooters that need to be sold in July through December to meet the sales goal? ____________

Lesson 1 Problem Solving

Use the bar graph to answer each question.

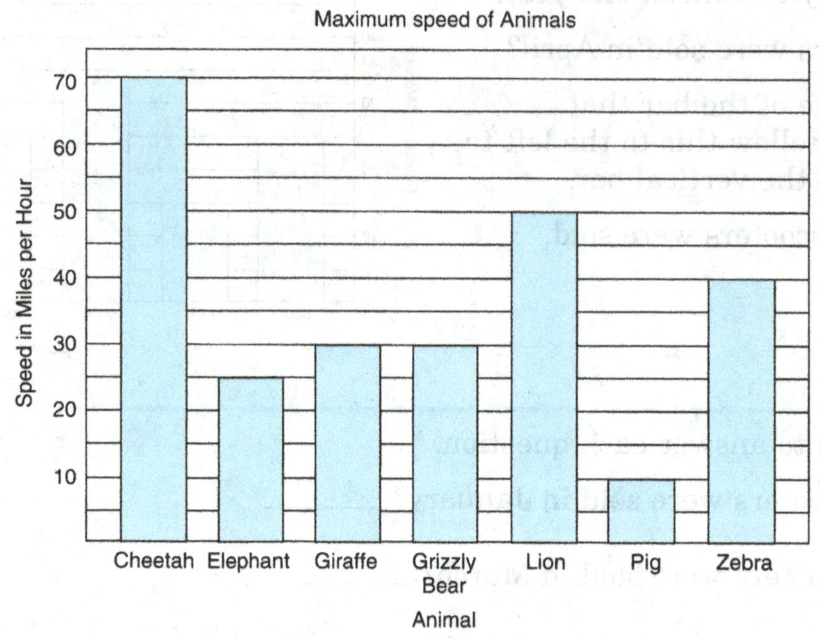

1. What is the maximum speed of a cheetah? ____________

2. What is the maximum speed of a giraffe? ____________

3. What is the maximum speed of a grizzly bear? ____________

4. What is the maximum speed of a pig? ____________

5. What is the maximum speed of a zebra? ____________

6. How much faster is a lion than an elephant? ____________

7. How much slower is a pig than a grizzly bear? ____________

8. How much faster is a cheetah than a zebra? ____________

9. Which animals are five miles per hour faster than an elephant? ____________

10. If an elephant, giraffe, and zebra were racing, which animal would most likely win? ____________

NAME ____________________

Lesson 2 Line Graphs

The **line graph** shows the distance Jordy rode on his bike ride Saturday.

How many miles did Jordy ride by 10:00 A.M.?

Locate the point on the graph that represents 10:00 A.M. Follow this to the left to locate its value on the vertical bar.

By 10:00 A.M., Jordy rode
__15__ miles.

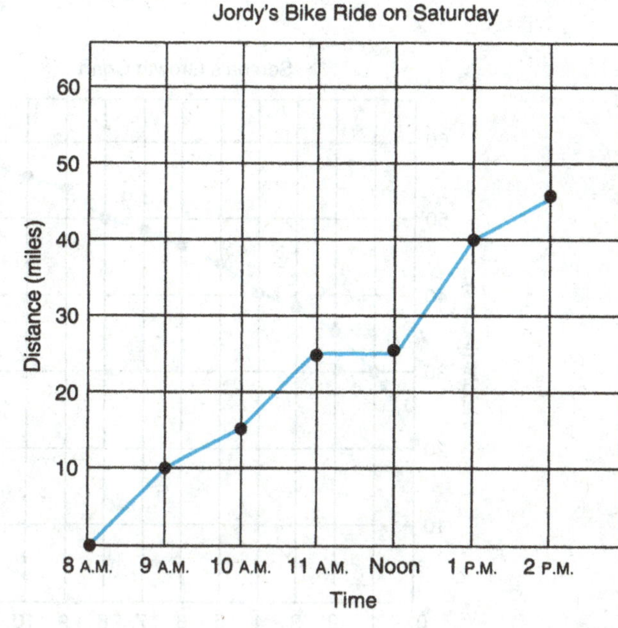

CHAPTER 7

Use the line graph of Jordy's bike ride to answer each question.

1. How many miles did Jordy ride by 8:00 A.M.? ____________

2. How many miles did Jordy ride by 11:00 A.M.? ____________

3. How many miles did Jordy ride by 12:00 noon? ____________

4. How many miles did Jordy ride by 2:00 P.M.? ____________

5. Between which two consecutive times graphed did Jordy ride the farthest distance? ____________

6. Between which two consecutive times graphed did Jordy ride the shortest distance? ____________

7. About how many miles did Jordy ride by 10:30 A.M.? ____________

8. How many more miles did Jordy bike between 12:00 noon and 1:00 P.M. than he biked between 1:00 P.M. and 2:00 P.M.? ____________

9. About what time do you think Jordy stopped to eat lunch? ____________

10. On Sunday, Jordy biked one-third as far as he did on Saturday. How many miles did Jordy bike on Sunday? ____________

Lesson 2 Problem Solving

Use the line graph to answer each question.

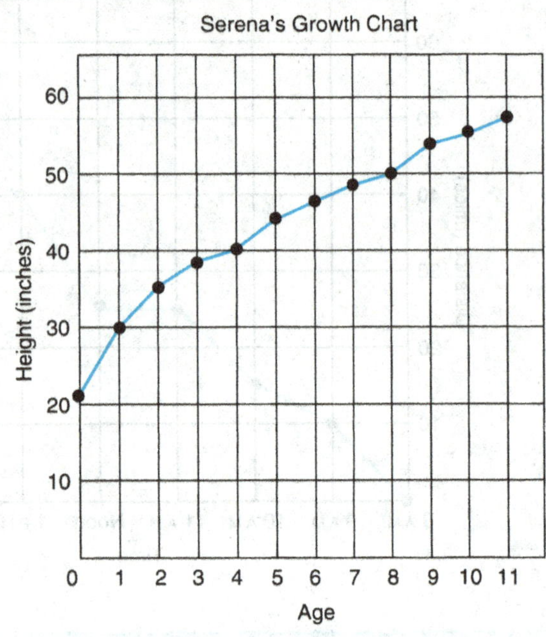

What was Serena's approximate height at the following ages?

1. 0/birth ______ age 2 ______ age 5 ______

2. age 6 ______ age 8 ______ age 11 ______

3. What is the difference in Serena's height at age 4 and her height at age 1? ______

4. What is the difference in Serena's height at age 10 and her height at age 4? ______

5. Between which two consecutive ages did Serena grow the most?
______

6. Between which two consecutive ages did Serena grow 5 inches?
______

7. If Serena grows three more inches by the time she is 12 years old, how tall will she be? ______

8. When Serena was 8 years old she was 16 inches shorter than her mom. How tall is Serena's mom? ______

Lesson 3 Mean

To find the **mean** or average of a set of numbers, add up all the numbers and then divide by the number of addends.

What is the mean of 6, 10, 11, 9?

Add the values.

$$\begin{array}{r} \scriptstyle 1 \\ 6 \\ 10 \\ 11 \\ +9 \\ \hline 36 \end{array}$$

Divide by 4, the number of addends.

$$\begin{array}{r} 9 \\ 4\overline{)36} \\ \underline{36} \\ 0 \end{array}$$

The mean of 6, 10, 11, 9 is ___9___.

CHAPTER 7

Find the mean of each set of numbers.

	a	*b*
1.	6, 9, 5, 8 ______	13, 21, 10, 16 ______
2.	12, 17, 19, 11, 16 ______	26, 22, 27, 30, 25 ______
3.	42 ft, 35 ft, 33 ft, 41 ft, 39 ft ______	88%, 97%, 92%, 95% ______
4.	\$2.48, \$2.67, \$2.28, \$2.45 ______	\$1.25, \$1.64, \$1.38, \$1.29 ______

Lesson 3 Problem Solving

Solve each problem.

1. Betsy bought 4 books at the bookstore. The prices of the books were \$4, \$5, \$8, and \$3. What is the mean price of the books?

   The mean price of the books is ________.

   1.

2. Keshawn kept track of how long it took him to drive to work every day this week. On Monday, it took 18 minutes. On Tuesday, it took 22 minutes. On Wednesday, it took 18 minutes. On Thursday, it took 21 minutes. On Friday, it took 26 minutes. What is the mean amount of time it took Keshawn to drive to work this week?

   The mean amount of time it took Keshawn to drive to work this week is ________ minutes.

   2.

3. On the first three math tests this quarter, Yoshiyo scored 87%, 95%, and 91%. What is the mean score of Yoshiyo's first three math tests?

   The mean score of Yoshiyo's first three math tests is ________.

   3.

4. Ryan and four of his friends measured their heights in inches. Their heights were 57 inches, 61 inches, 58 inches, 57 inches, and 62 inches. What is the mean height of Ryan and his friends?

   The mean height of Ryan and his friends is ________ inches.

   4.

5. Jorge bought his lunch three days this week at school. He spent \$2.50 on Monday, \$3.25 on Tuesday, and \$2.80 on Thursday. What is the mean amount of money that Jorge spent on lunch this week?

   The mean amount of money that Jorge spent on lunch this week is ________.

   5.

NAME ______________________

Lesson 4 Median, Mode, and Range

The **median** is the middle number of a set of numbers.

The **mode** is the number that appears most often in the set of numbers.

The **range** is the difference between the greatest and least number in the set.

What is the median, mode, and range of 9, 14, 11, 7, 9?

Order the numbers from least to greatest to find the median.

7, 9, 9, 11, 14 The median is 9.

The number 9 appears most often. The mode is 9.

Subtract 7 (least) from 14 (greatest) to find the range.

14 − 7 = 7 The range is 7.

CHAPTER 7

Find the median, mode, and range of each set of numbers.

	a		*b*	
1.	5, 9, 5, 10, 6	median: ______ mode: ______ range: ______	9, 3, 6, 4, 8, 3, 8, 4, 3	median: ______ mode: ______ range: ______
2.	15, 17, 11, 13, 15, 10, 14	median: ______ mode: ______ range: ______	33, 37, 30, 29, 34, 37, 39	median: ______ mode: ______ range: ______
3.	87%, 80%, 76%, 84%, 80%	median: ______ mode: ______ range: ______	67%, 78%, 85%, 77%, 77%, 73%, 89%	median: ______ mode: ______ range: ______
4.	$127, $105, $113, $120, $127	median: ______ mode: ______ range: ______	$58, $65, $44, $63, $44, $65, $44	median: ______ mode: ______ range: ______

Lesson 4 Problem Solving

Answer each question.

1. Shanelle ran five times this week. On Monday, she ran 3 miles. On Wednesday, she ran 5 miles. On Thursday, she ran 3 miles. On Friday, she ran 4 miles. On Saturday, she ran 6 miles.

   What is the median of the number of miles Shanelle ran this week? ________

   What is the mode of the number of miles Shanelle ran this week? ________

   What is the range of the number of miles Shanelle ran this week? ________

2. David kept track of the amounts of his phone bills over the past five months. The amounts of the phone bills were $22, $16, $34, $28, and $16.

   What is the median amount of the phone bills? ________

   What is the mode amount of the phone bills? ________

   What is the range of the phone bills? ________

3. Enrique had nine math tests all school year. His scores were 78%, 87%, 70%, 76%, 88%, 92%, 73%, 88%, and 90%.

   What is the median score of Enrique's math tests? ________

   What is the mode score of Enrique's math tests? ________

   What is the range of the scores of Enrique's math tests? ________

4. An electronic store carries five different 27-inch televisions. The prices of these televisions are $325, $395, $405, $485, and $325.

   What is the median price of televisions at the electronic store? ________

   What is the mode price of televisions at the electronic store? ________

   What is the range of the television prices at the electronic store? ________

1.

2.

3.

4.

Lesson 5 Probabilities

Probability is the chance that something will occur.

What is the probability of spinning a 3?

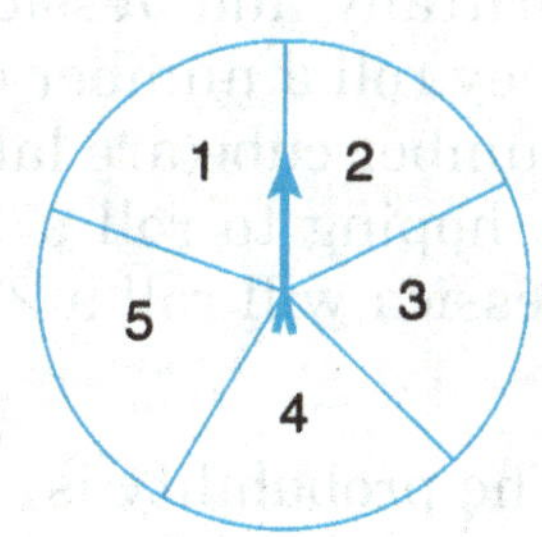

Spinning a 3 is the favorable outcome. There are 5 possible outcomes.

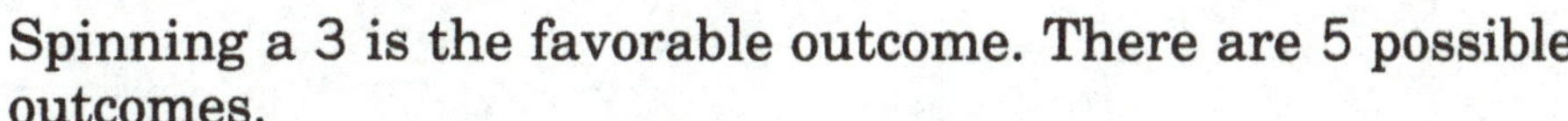

number or possible outcomes → 5

The probability of spinning a 3 is $\frac{1}{5}$.

You spin the wheel shown at the right. Find the probability of the spinner stopping on:

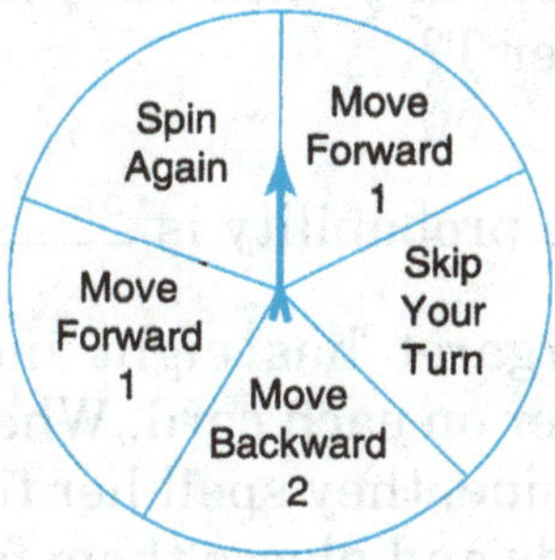

1. Skip Your Turn ______________

2. Move Forward 1 ______________

3. Move Backward 2 ______________

4. Spin Again ______________

A bag contains 11 marbles. Five marbles are yellow. Three marbles are red. Two marbles are green. One marble is blue. What is the probability of selecting a marble of the given color without looking?

	a	*b*
5.	red ______________	yellow ______________
6.	blue ______________	green ______________
7.	yellow and blue ______________	red and green ______________

Lesson 5 Problem Solving

Find each probability.

1. Brittany and Jessica are playing a game in which they roll a number cube each turn. The sides of the number cube are labeled 1–6. On Jessica's turn she is hoping to roll a 2. What is the probability that Jessica will roll a 2?

   The probability is ______.

2. Damian has one $10 bill, three $5 bills, and six $1 bills in his wallet. If he selects one bill from his wallet without looking, what is the probability that he will select a $5 bill?

   The probability is ______.

3. Luis wrote each of the letters of the alphabet on a separate card. He placed all of these cards face down on the floor. If he selects one card, what is the probability that he will select the card with the letter T?

   The probability is ______.

4. Margaret has eight index cards. She writes one letter on each card. When she places the cards side-by-side, they spell her first name. She mixes up the cards and places them face down on her desk. If she selects one of the cards, what is the probability that it will have the letter R or T on it?

   The probability is ______.

5. The Whitman family just bought a new dog. They were trying to think of a name for the dog. Their ideas were Fluffy, Baxter, Patch, Babe, and Bailey. They wrote all their ideas on separate pieces of paper and placed them in a bowl. Without looking, they selected the name from the bowl. What is the probability that the paper they selected was Patch?

   The probability is ______.

1.
2.
3.
4.
5.

NAME ____________________

CHAPTER 7 PRACTICE TEST
Graphs and Averages

Use the bar graph to answer each question.

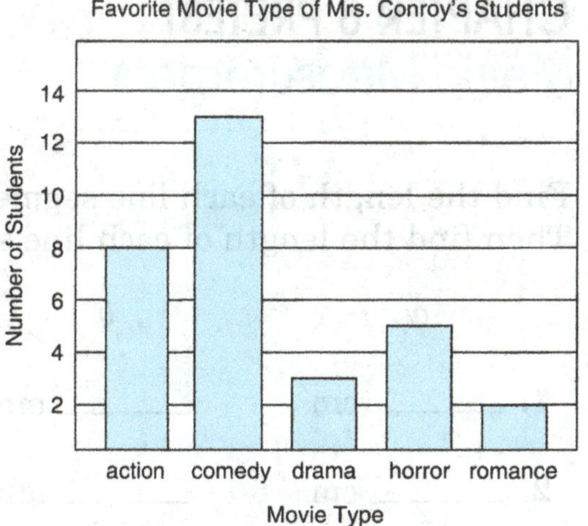

1. How many students chose comedy as their favorite movie type? ________

2. How many students chose action as their favorite movie type? ________

3. How many more students chose horror as their favorite movie type than romance? ________

Use the line graph to answer each question.

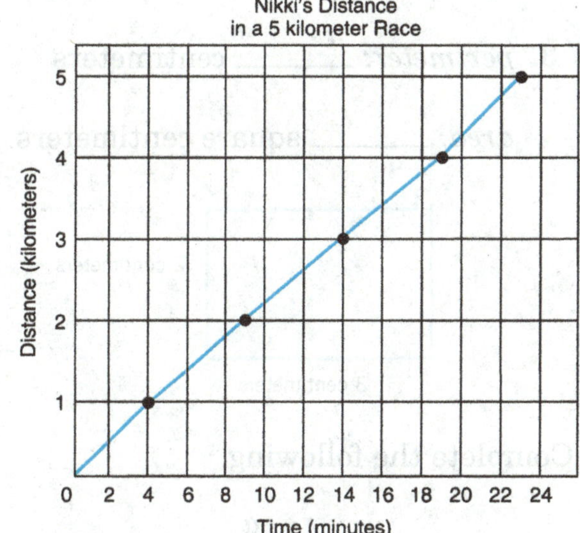

4. How long did it take Nikki to run the first kilometer? ____________

5. After 14 minutes, how far did Nikki run? ______________

6. How long did it take Nikki to run 4 kilometers? ______________

7. What is the quickest amount of time Nikki ran a kilometer in the race? ____________

Find the mean, median, mode, and range of each set of numbers.

	a	*b*
8.	78%, 84%, 87%, 95%, 83%, 77%, 84%	\$125, \$164, \$118, \$114, \$164
	mean: ________	mean: ________
	median: ________	median: ________
	mode: ________	mode: ________
	range: ________	range: ________

9. A bag contains eight marbles. Four marbles are black. Three marbles are white. One marble is red. What is the probability of selecting a white marble without looking? ________

NAME ______________________

CHAPTER 8 PRETEST
Metric Measurement

Find the length of each line segment to the nearest centimeter (cm).
Then find the length of each line segment to the nearest millimeter (mm).

	a	*b*
1.	_______ cm	_______ mm
2.	_______ cm	_______ mm

Find the perimeter and the area of each rectangle.

3. *perimeter:* _______ centimeters

area: _______ square centimeters

4. *perimeter:* _______ millimeters

area: _______ square millimeters

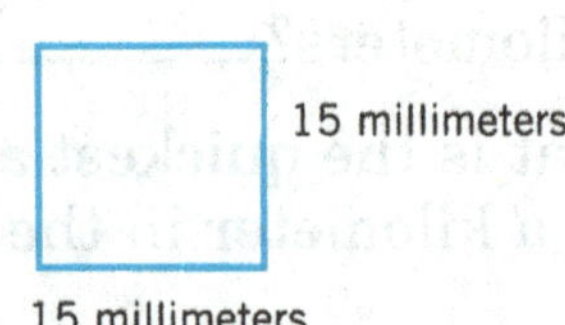

Complete the following.

	a	*b*
5.	7 centimeters = _______ millimeters	28 meters = _______ centimeters
6.	9 meters = _______ centimeters	49 meters = _______ millimeters
7.	8 kilometers = _______ meters	16 liters = _______ milliliters
8.	5 kiloliters = _______ liters	5 kilograms = _______ grams
9.	2 grams = _______ milligrams	14 centimeters = _______ millimeters
10.	40 liters = _______ milliliters	42 meters = _______ centimeters
11.	3 kiloliters = _______ liters	35 meters = _______ millimeters
12.	60 kilograms = _______ grams	34 kilometers = _______ meters

NAME ____________________

Lesson 1 Centimeter and Millimeter

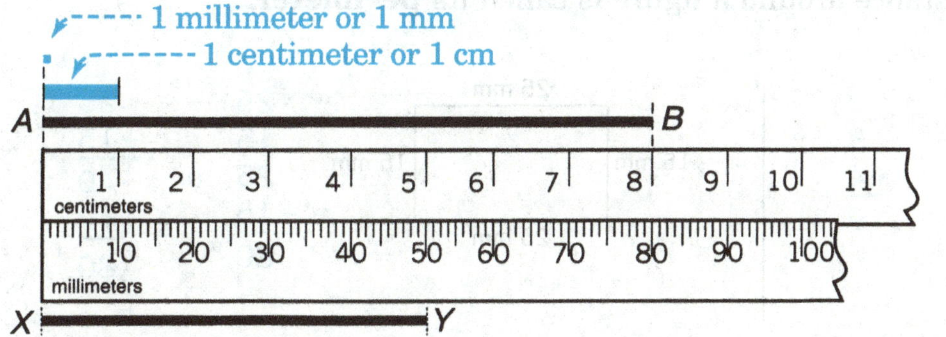

Line segment *AB* is ___8___ centimeters long. *XY* is ______ centimeters long.

Line segment *AB* is ___80___ millimeters long. *XY* is ______ millimeters long.

CHAPTER 8

Find the length of each line segment to the nearest centimeter.
Then find the length of each line segment to the nearest millimeter.

a *b*

1. ______ cm ______ mm

2. ______ cm ______ mm

3. ______ cm ______ mm

4. ______ cm ______ mm

Find the length of each line segment to the nearest millimeter.

5. ______ mm

6. ______ mm

7. ______ mm

8. ______ mm

Draw a line segment for each measurement.

9. 6 cm

10. 45 mm

NAME ______________________

Lesson 2 Perimeter

The distance around a figure is called its perimeter.

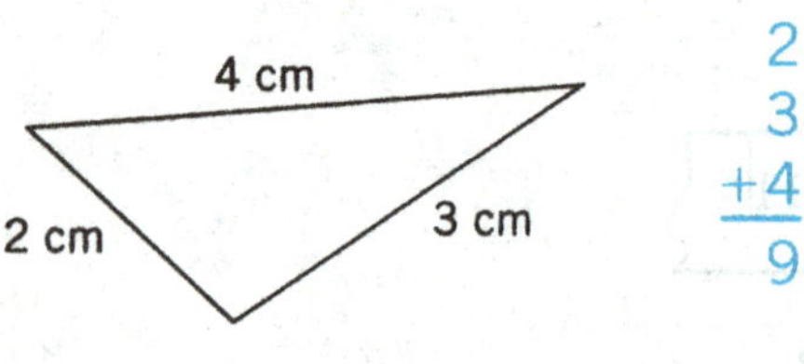

$$\begin{array}{r} 2 \\ 3 \\ +4 \\ \hline 9 \end{array}$$

perimeter: 9 cm

25 mm
15 mm | 15 mm
25 mm

$$\begin{array}{r} 25 \\ 15 \\ 25 \\ +15 \\ \hline 80 \end{array} \quad \text{or} \quad \begin{array}{r} 25 \\ +15 \\ \hline 40 \\ \times 2 \\ \hline 80 \end{array}$$

perimeter: 80 mm

Measure each side in centimeters. Then find the perimeter of each figure.

	a	*b*
1.	________ cm	________ cm

	a	*b*
2.	________ cm	________ cm

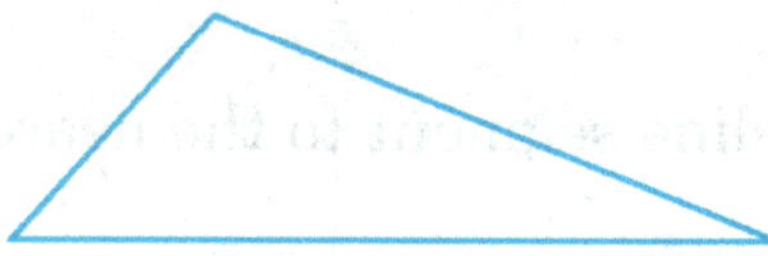

Measure each side in millimeters. Then find the perimeter of each figure.

	a	*b*
3.	________ mm	________ mm

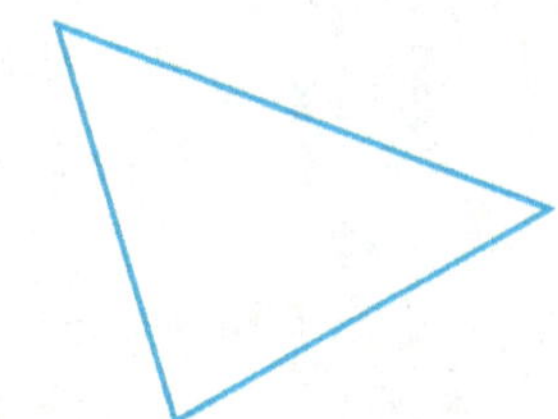

NAME ______________________

Lesson 3 Meter and Kilometer

A baseball bat is about **1 meter** long.

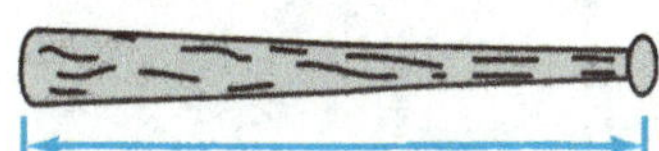

1 meter (m) or 100 cm

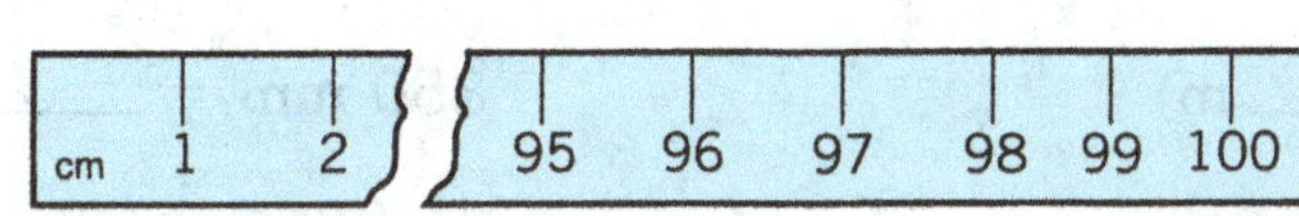

1 m = 100 cm
1 cm = 0.01 m

If you run from goal line to goal line on a football field **11** times, you will run about **1 kilometer.**

1,000 meters is the same distance as **1** kilometer (km).

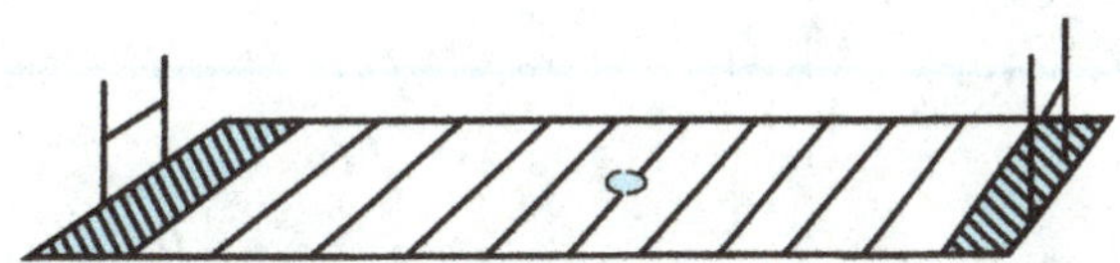

1 km = 1,000 m
1 m = 0.001 km

Use a meterstick to find the following to the nearest meter.

	a	*b*
1.	length of your room ______ m	width of a door ______ m
2.	width of your room ______ m	width of a window______ m
3.	height of a door ______ m	height of a window ______ m

Answer each question.

4. Michelle's height is 105 centimeters. Is she taller or shorter than 1 meter?

She is ________ than 1 meter.

4.

5. Are you taller or shorter than 1 meter?

I am ________ than 1 meter.

5.

6. Roberta wants to swim 1 kilometer. How many meters should she swim?

She should swim ________ meters.

6.

7. Sung-Chi ran 1,500 meters. Leona ran 1 kilometer. Who ran farther? How much farther?

__________ ran __________ meters farther.

7.

CHAPTER 8

NAME ________________

Lesson 4 Units of Length

Study how to change from one metric unit to another.

9 km = __?__ m

1 km = 1,000 m

9 km = (9 × 1000) m

9 km = __9,000__ m

850 mm = __?__ cm

10 mm = 1 cm

850 mm = (850 ÷ 10) cm

850 mm = __85__ cm

Complete the following.

	a	*b*
1.	50 km = ________ m	600 cm = ________ m
2.	70 mm = ________ cm	2,000 mm = ________ m
3.	9 cm = ________ mm	8,000 m = ________ km
4.	3 m = ________ cm	5,000 cm = ________ m

5. Ted is 4,000 meters from school. Susan is 3 kilometers from school. How many meters from school is Susan? Who is farther from school? How much farther?

Susan is ________ meters from school.

________ is ________ meters farther from school.

5.

6. Maria is 134 centimeters tall. Su-Lyn is 1,300 millimeters tall. Charles is 141 centimeters tall. Who is tallest? Who is shortest?

________ is tallest.

________ is shortest.

6.

7. What is your height in centimeters? In millimeters?

I am ________ centimeters tall.

I am ________ millimeters tall.

7.

NAME ______________________

Lesson 5 Area

To find the **area** of a rectangle, multiply the measure of its length by the measure of its width.

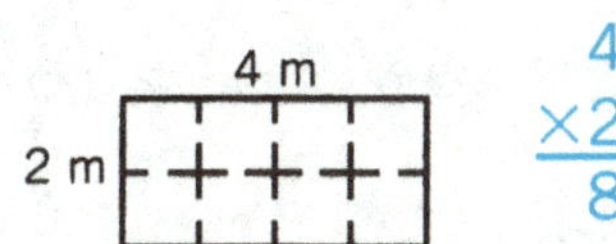

$$\begin{array}{r} 4 \\ \times 2 \\ \hline 8 \end{array}$$

area: 8 square meters

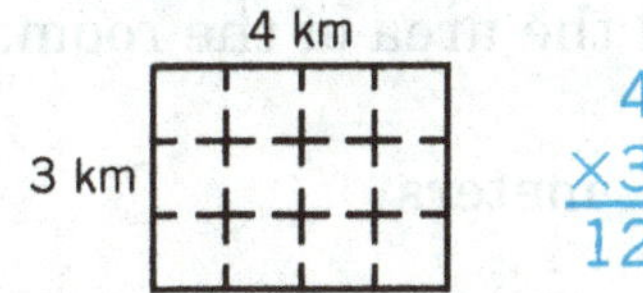

$$\begin{array}{r} 4 \\ \times 3 \\ \hline 12 \end{array}$$

area: 12 square kilometers

Find the area of each rectangle.

	a	*b*	*c*
1.	______ square kilometers	______ square millimeters	______ square meters

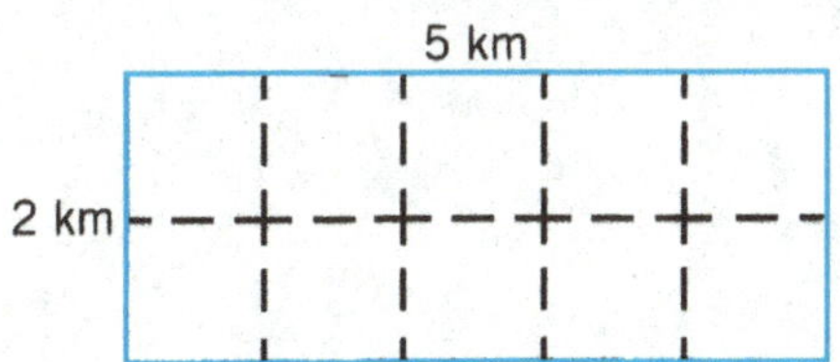

60 mm

50 mm

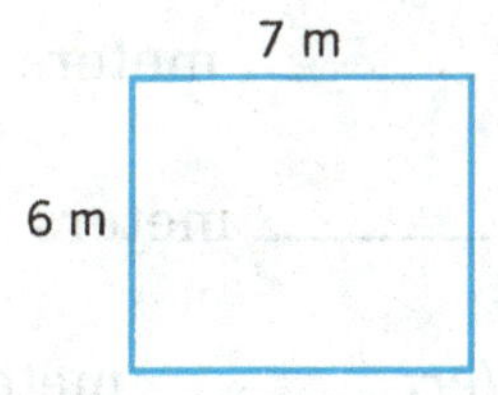

	a	*b*	*c*
2.	______ square meters	______ square kilometers	______ square centimeters

35 m

35 m

27 km

20 km

15 cm

10 cm

	length	*width*	*area*
3.	9 km	6 km	______ square kilometers
4.	18 cm	7 cm	______ square centimeters
5.	14 m	10 m	______ square meters
6.	175 mm	25 mm	______ square millimeters
7.	152 cm	100 cm	______ square centimeters

CHAPTER 8

Lesson 5 Problem Solving

Solve each problem.

1. Find a rectangular room. Measure its length and width to the nearest meter. Find the perimeter of the room. Find the area of the room.

   length: ______ meters

   width: ______ meters

   perimeter: ______ meters

   area: ______ square meters

2. Find a rectangular tabletop or desk. Measure its length and width to the nearest meter. Find the perimeter of the top. Find the area of the top.

   length: ______ meters

   width: ______ meters

   perimeter: ______ meters

   area: ______ square meters

3. Use the front cover of this book. Measure its length and width to the nearest centimeter. Find the perimeter of the cover. Find the area of the front cover.

   perimeter: ______ centimeters

   area: ______ square centimeters

4. Use the rectangle at the right. Find the perimeter of the rectangle. Find the area of the rectangle.

   perimeter: ______ millimeters

   area: ______ square millimeters

35 mm

24 mm

1.

2.

3.

4.

NAME ______________________

Lesson 6 Volume

To find the **volume** of a rectangular solid, multiply the measure of its length by the measure of its width by the measure of its height.

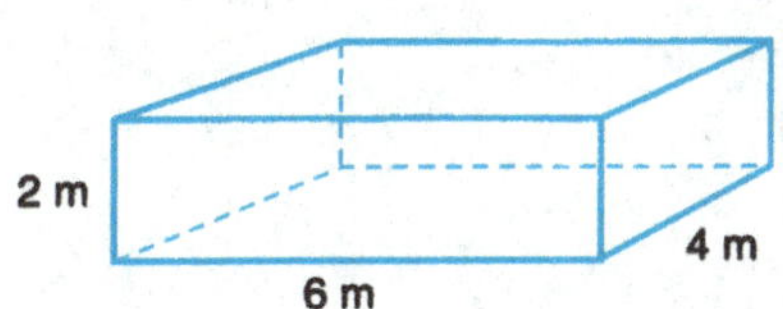

$$\begin{array}{r} 6 \\ \times 4 \\ \hline 24 \\ \times 2 \\ \hline 48 \end{array}$$

volume: 48 cubic meters

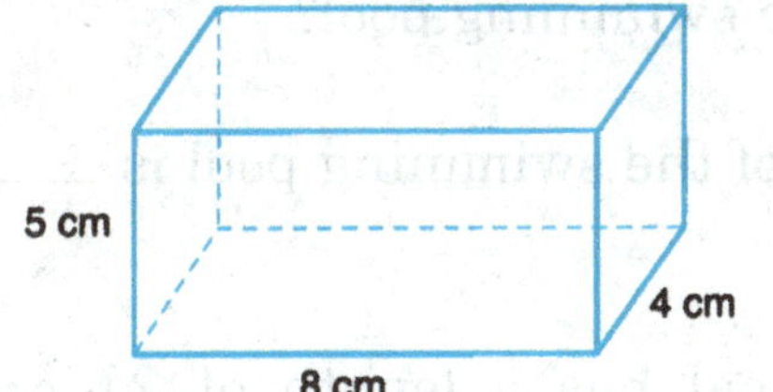

$$\begin{array}{r} 8 \\ \times 4 \\ \hline 32 \\ \times 5 \\ \hline 160 \end{array}$$

volume: 160 cubic centimeters

Find the volume of each rectangle.

	a	*b*	*c*
1.	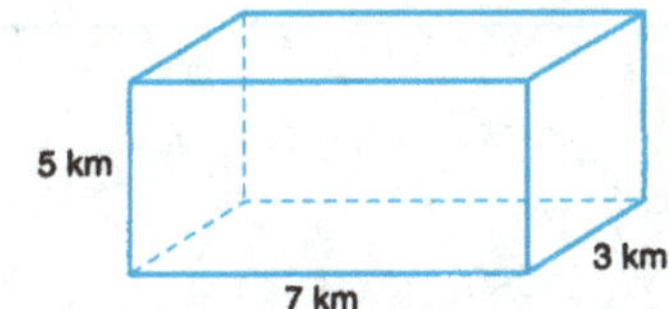 5 km, 7 km, 3 km ______ cubic kilometers	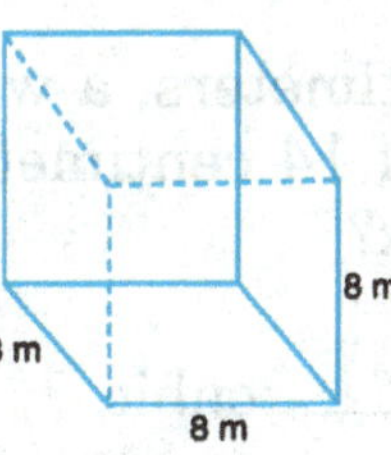 8 m, 8 m, 8 m ______ cubic meters	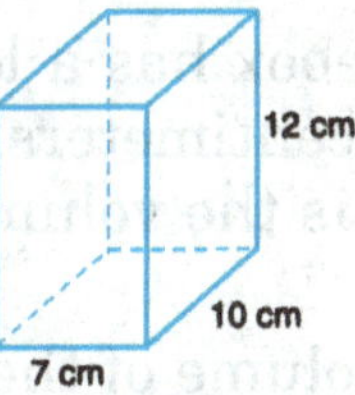 12 cm, 10 cm, 7 cm ______ cubic centimeters
2.	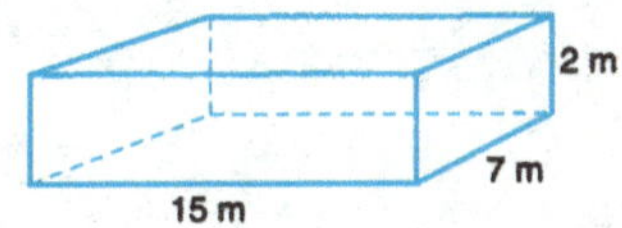 2 m, 15 m, 7 m ______ cubic meters	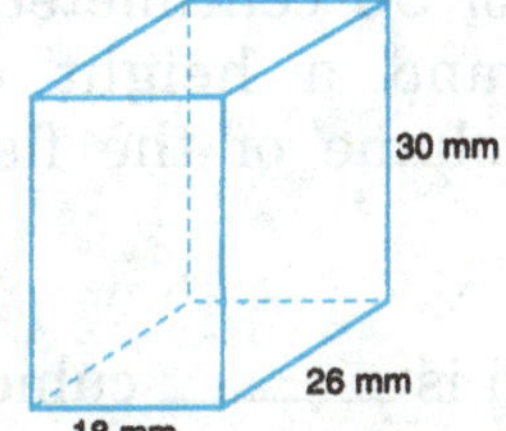 30 mm, 26 mm, 18 mm ______ cubic millimeters	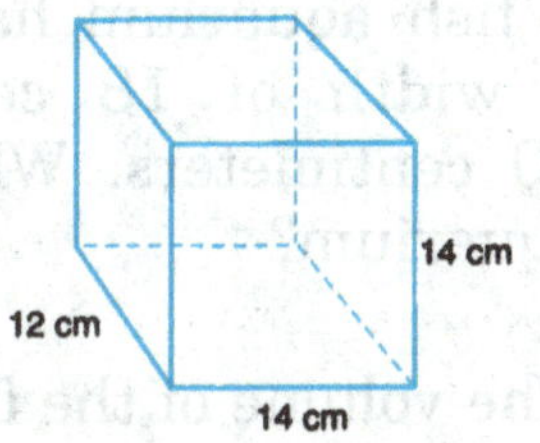 14 cm, 12 cm, 14 cm ______ cubic centimeters

	length	*width*	*height*	*volume*
3.	5 m	6 m	7 m	______ cubic meters
4.	9 km	3 km	8 km	______ cubic kilometers
5.	10 cm	13 cm	6 cm	______ cubic centimeters
6.	26 mm	32 mm	15 mm	______ cubic millimeters

CHAPTER 8

Lesson 6 Problem Solving

Solve each problem.

1. A swimming pool has a length of 7 meters, a width of 4 meters, and a depth of 2 meters. What is the volume of the swimming pool?

   The volume of the swimming pool is ______ cubic meters.

   1.

2. A box of cereal has a length of 21 centimeters, a width of 6 centimeters, and a height of 30 centimeters. What is the volume of the cereal box?

   The volume of the cereal box is ______ cubic centimeters.

   2.

3. A shoebox has a length of 28 centimeters, a width of 20 centimeters, and a height of 14 centimeters. What is the volume of the shoebox?

   The volume of the shoebox is ______ cubic centimeters.

   3.

4. A fish aquarium has a length of 36 centimeters, a width of 18 centimeters, and a height of 20 centimeters. What is the volume of the fish aquarium?

   The volume of the fish aquarium is ______ cubic centimeters.

   4.

5. Find a shoebox that is a rectangular solid. Measure its length, width, and height. Find the volume of the shoebox.

   length: ______________________

   width: ______________________

   height: ______________________

   volume: ______________________

   5.

NAME ____________

Lesson 7 Capacity

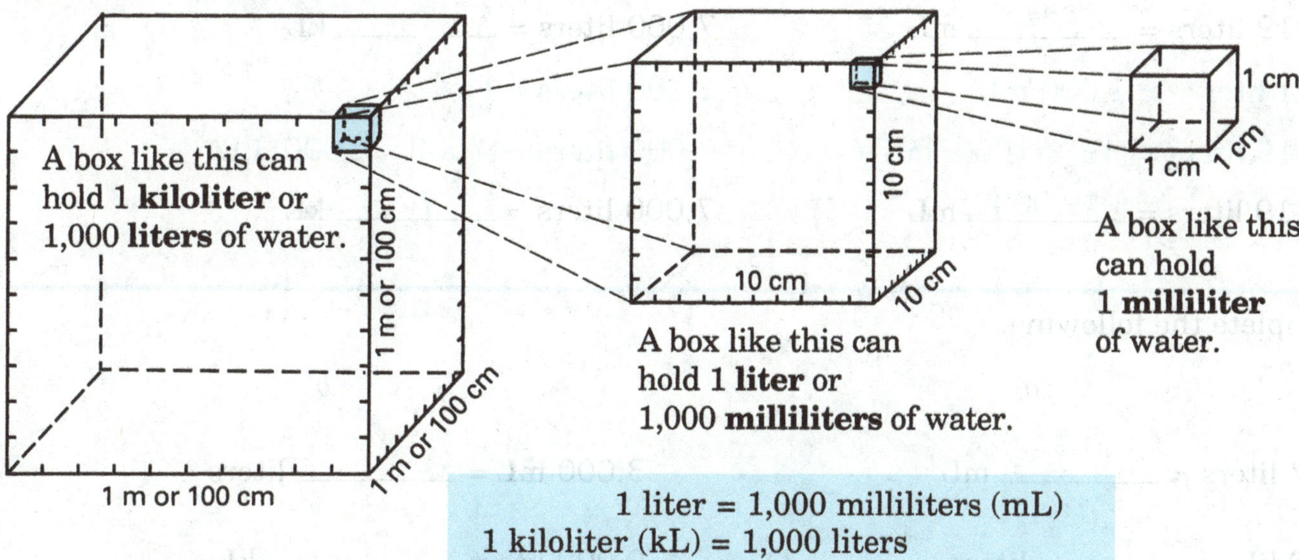

1 liter = 1,000 milliliters (mL)
1 kiloliter (kL) = 1,000 liters

Solve each problem.

1. A teaspoon holds about 5 milliliters. A recipe calls for 2 teaspoons of vanilla. How many milliliters is that?

That is ______ milliliters.

1.

2. A liter is slightly more than 4 cups. Do you drink more or less than a liter of milk every day?

I drink ______________ than a liter every day.

2.

3. To make punch, 8 cups of fruit juice are used. About how many liters would that be?

That would be ______ liters.

3.

4. Two bathtubs filled with water would be about 1 kiloliter of water. Suppose your family uses 10 tubfuls of water a week. How many kiloliters of water would be used in a week?

______ kiloliters would be used in a week.

4.

5. A tank holds 1,000 liters. How many kiloliters would it hold?

It would hold ______ kiloliter(s).

5.

CHAPTER 8

NAME ______________________

Lesson 8 Units of Capacity

19 liters = ___?___ mL	7,000 liters = ___?___ kL
1 liter = 1,000 mL	1,000 liters = 1 kL
19 liters = (19 × 1000) mL	7,000 liters = (7000 ÷ 1000) kL
19 liters = __19,000__ mL	7,000 liters = ___7___ kL

Complete the following.

	a	*b*
1.	7 liters = ________ mL	3,000 mL = ________ liters
2.	2 kL = ________ liters	9,000 liters = ________ kL
3.	20 liters = ________ mL	48 kL = ________ liters
4.	4,000 mL = ________ liters	5,000 liters = ________ kL

5. Lisa filled an ice-cube tray with water. Do you think she used about 1 *milliliter*, 1 *liter*, or 1 *kiloliter* of water?

She used 1 ________ of water.

5.

6. Carlos said he drank 500 milliliters of milk. Larry said he drank 1 liter of milk. Who drank more milk? How many milliliters more?

________ drank ________ milliliters more milk.

6.

7. The gasoline tank on Mrs. Mohr's car holds 85 liters. It took 27 liters of fuel to fill the tank. How much fuel was in the tank before it was filled?

________ liters were in the tank.

7.

8. A tank can hold 4,000 liters of water. There are 3 kiloliters of water in the tank. How many liters of water are needed to fill the tank?

________ liters are needed.

8.

Lesson 9 Weight

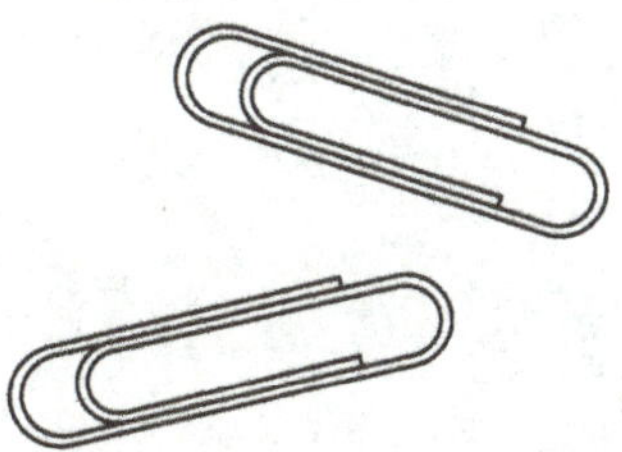

2 paper clips weigh about 1 **gram** (g).

3 math books like yours weigh about 1 **kilogram** (kg).

1 gram = 1,000 milligrams (mg)
1,000 grams = 1 kilogram (kg)

Complete the following.

1. About how many grams do four paper clips weigh? | **1.**

They weigh about ________ grams.

2. A box contains 4,000 paper clips. How many kilograms do those paper clips weigh? | **2.**

They weigh ________ kilograms.

3. One nickel weighs about 5 grams. A roll of 40 nickels would weigh about how many grams? | **3.**

It would weigh ________ grams.

4. How many kilograms would six math books like yours weigh? | **4.**

They would weigh ________ kilograms.

5. A doctor has 3,000 milligrams of medicine. How many grams is that? | **5.**

That is ________ grams.

6. A dog weighs 17,000 grams. How many kilograms is that? | **6.**

That is ________ kilograms.

NAME ____________________

Lesson 10 Units of Weight

6 kg = ___?___ g

1 kg = 1,000 g

6 kg = (6 × 1000) g

6 kg = ___6,000___ g

5,000 mg = ___?___ g

1,000 mg = 1 g

5,000 mg = (5000 ÷ 1000) g

5,000 mg = ___5___ g

Complete the following.

	a	*b*
1.	2 kg = ________ g	6 g = ________ mg
2.	9 g = ________ mg	9 kg = ________ g
3.	2,000 mg = ________ g	7,000 g = ________ kg
4.	3,000 g = ________ kg	8,000 mg = ________ g

5. A penny weighs about 3 grams. A dime weighs about 2,000 milligrams. Which weighs more? How much more?

A ________ weighs about ________ milligrams more.

5.

6. Emily uses a 4-kilogram bowling ball. Her father uses a 7-kilogram bowling ball. How much heavier is her father's bowling ball?

It is ________ kilograms heavier.

6.

7. A loaf of bread weighs 454 grams. How much would 3 loaves of bread weigh?

They would weigh ________ grams.

7.

8. John weighs 34,000 grams. Judy weighs 39 kilograms. Who weighs more? How much more?

________ weighs ________ kilograms more.

8.

NAME ______________________

CHAPTER 8 PRACTICE TEST

Find the length of each line segment to the nearest centimeter.
Then find the length of each line segment to the nearest millimeter.

a *b*

1. _______ cm _______ mm

2. _______ cm _______ mm

Find the perimeter and the area of each rectangle.

3. *perimeter:* _______ meters

area: _______ square meters

6 m

4 m

4. *perimeter:* _______ millimeters

area: _______ square millimeters

25 mm

15 mm

Find the volume.

	length	*width*	*height*	*volume*
5.	9 cm	12 cm	6 cm	_______ cubic centimeters
6.	2 m	7 m	36 m	_______ cubic meters

Complete the following.

	a	*b*
7.	5 cm = _______ mm	2,000 m = _______ km
8.	700 cm = _______ m	300 mm = _______ cm
9.	6 km = _______ m	3 m = _______ cm
10.	4 kL = _______ liters	3,000 mL = _______ liters

CHAPTER 8

NAME ____________________

CHAPTER 9 PRETEST
Customary Measurement

Complete.

	a	*b*
1.	4 feet = ______ inches	4 feet 6 inches = ______ inches
2.	24 feet = ______ yards	2 yards 2 feet = ______ feet
3.	5 yards = ______ feet	1 yard 10 inches = ______ inches
4.	1 mile = ______ feet	8 feet 4 inches = ______ inches
5.	6 cups = ______ pints	3 quarts 1 pint = ______ pints
6.	8 quarts = ______ gallons	4 gallons 2 quarts = ______ quarts
7.	32 ounces = ______ pounds	6 pounds 6 ounces = ______ ounces

Find the perimeter of each figure.

	a	*b*	*c*
8.	______ feet	______ inches	______ yards

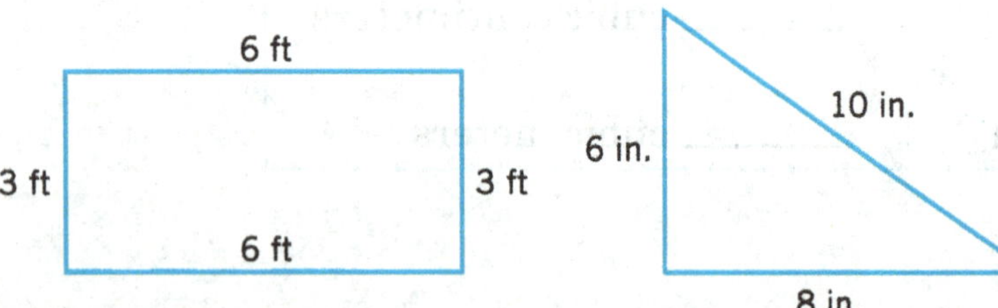

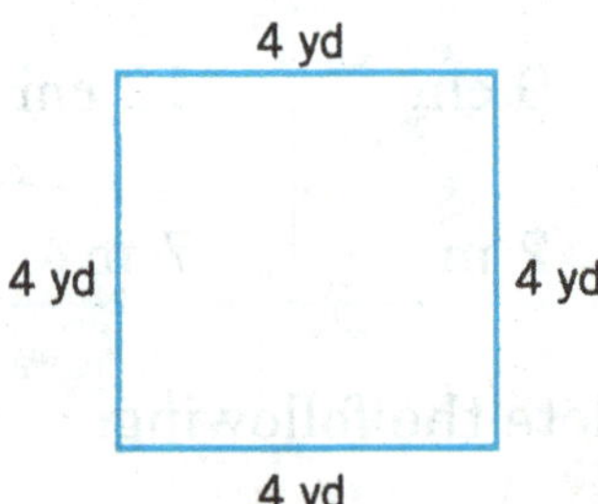

Find the area of each rectangle.

9.	______ square yards	______ square feet	______ square inches

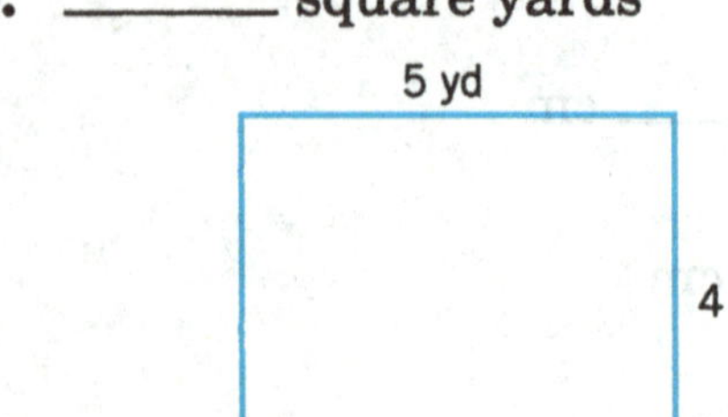

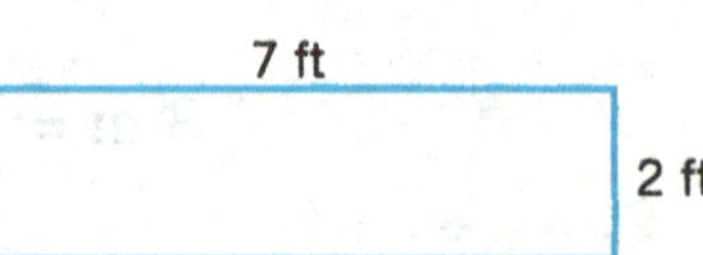

5 in.

5 in.

NAME ______________________

Lesson 1 Units of Length

1 foot (ft) = 12 inches (in.) 1 mile (mi) = 5,280 feet (ft)	1 yard (yd) = 3 ft 1 yard (yd) = 36 in.

24 in. = __?__ ft

12 in. = 1 ft

24 in. = (24 ÷ 12) ft

24 in. = __2__ ft

3 ft 4 in. = __?__ in.

1 ft = 12 in.

3 ft = (3 × 12) or 36 in.

3 ft 4 in. = 36 in. + 4 in.

3 ft 4 in. = ______ in.

Complete the following.

	a	*b*
1.	6 ft = ______ in.	3 ft 2 in. = ______ in.
2.	2 yd = ______ in.	6 yd 11 in. = ______ in.
3.	3 mi = ______ ft	1 mi 450 ft = ______ ft
4.	84 in. = ______ ft	7 yd 1 ft = ______ ft
5.	180 in. = ______ yd	4 yd 7 in. = ______ in.
6.	15 ft = ______ yd	2 ft 6 in. = ______ in.

7. Becky threw the ball 24 yards. Zachary threw the ball 840 inches. How many feet did each person throw the ball? Who threw it farther? How much farther?

Becky threw the ball ______ feet.

Zachary threw the ball ______ feet.

______ threw the ball ______ feet farther.

CHAPTER 9

Lesson 2 Perimeter

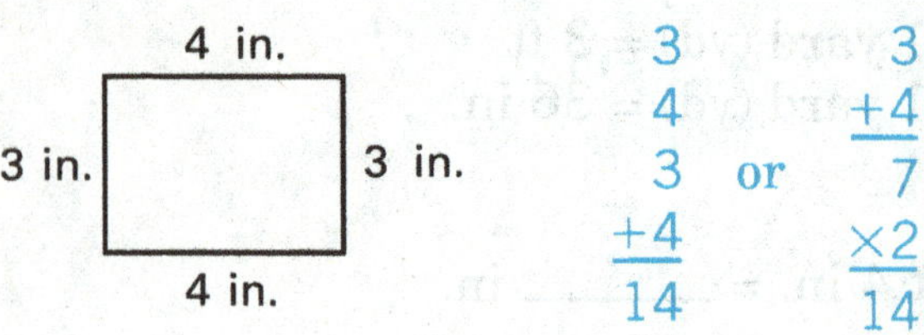

perimeter: 14 in.

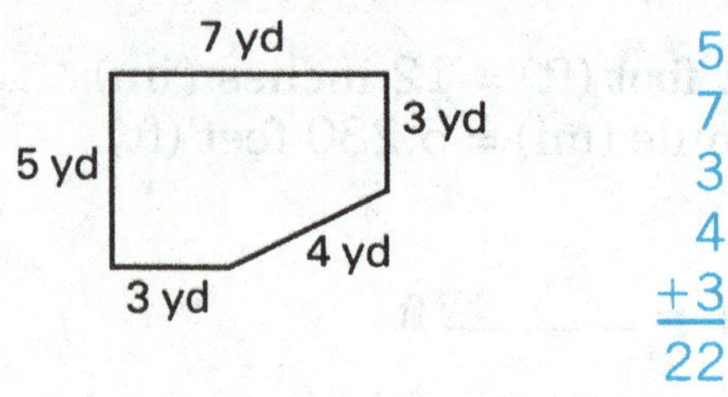

perimeter: ______ yd

Find the perimeter of each figure.

a | *b*

1. ______ inches

5 in.
4 in.
4 in.
5 in.

______ yards

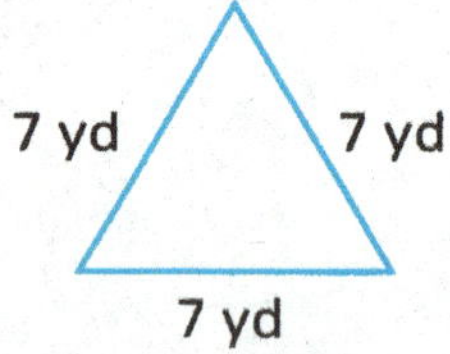

2. ______ feet

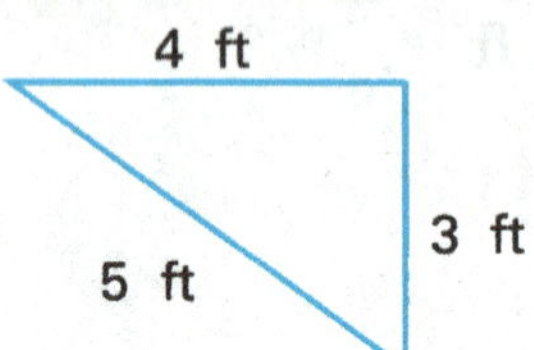

______ inches

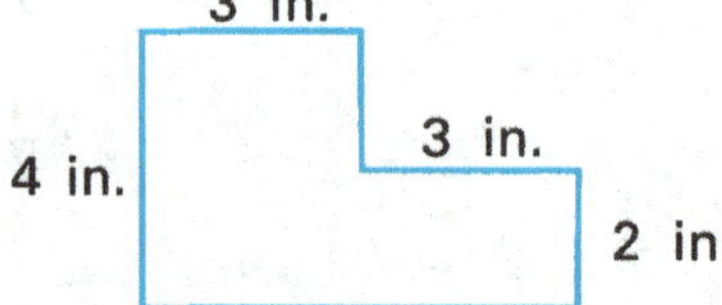

3. ______ inches

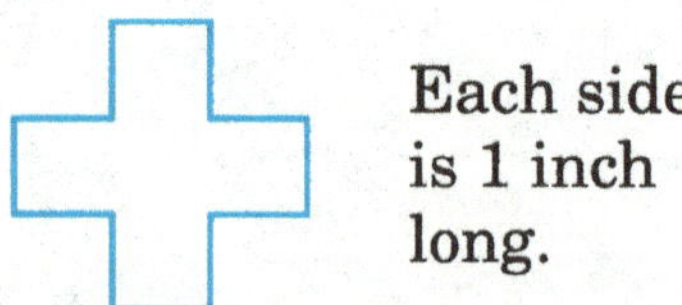

Each side is 1 inch long.

______ feet

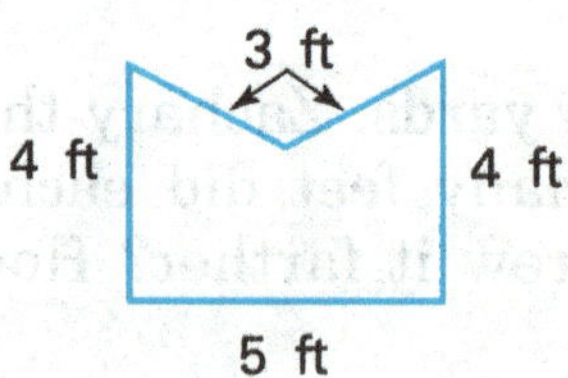

4. ______ yards

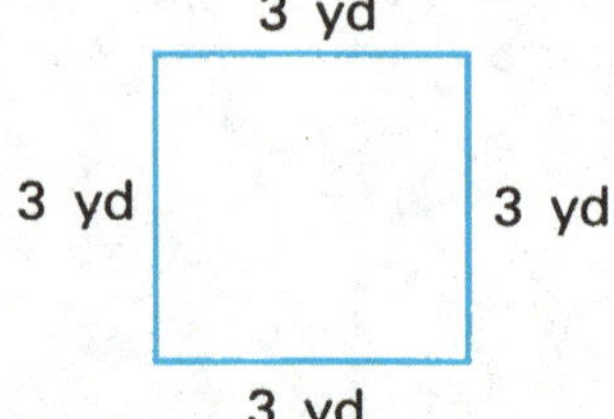

______ inches

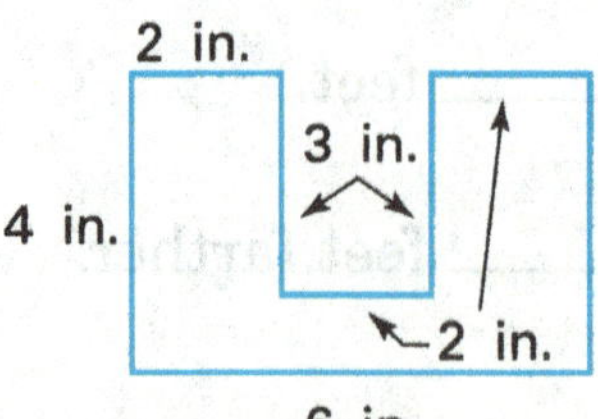

Lesson 3 Area

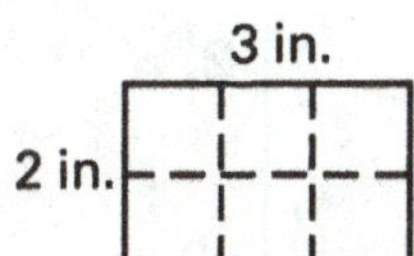

3
×2
6

area: 6 square inches

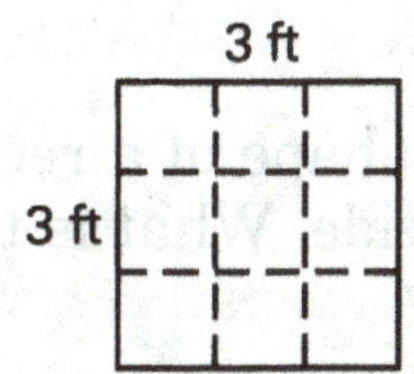

3
×3
9

area: 9 square feet

Find the area of each rectangle.

a | *b*

1. _______ square inches

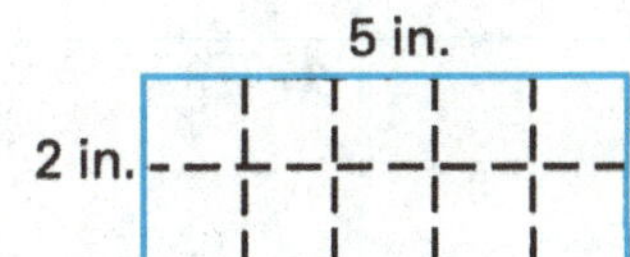

_______ square feet

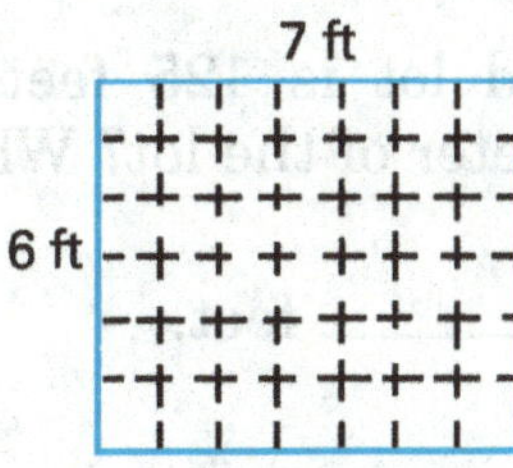

2. _______ square miles

8 mi
8 mi

_______ square feet

2 ft
2 ft

	length	*width*	*area*
3.	8 ft	5 ft	_______ square feet
4.	12 in.	8 in.	_______ square inches
5.	142 ft	57 ft	_______ square feet
6.	36 yd	12 yd	_______ square yards
7.	18 in.	15 in.	_______ square inches

Lesson 3 Problem Solving

Solve each problem.

1. A garden has the shape of a rectangle. It is 24 feet long and 10 feet wide. What is the perimeter of the garden?

   The perimeter is _______ feet.

2. A baseball diamond is a square with each side 90 feet long. Find the perimeter and the area of the diamond.

   The perimeter is _______ feet.

   The area is _______ square feet.

3. The square-shaped lot is 125 feet on each side. What is the perimeter of the lot? What is the area?

   The perimeter is _______ feet.

   The area is _______ square feet.

4. Find the perimeter and the area of the following figure.

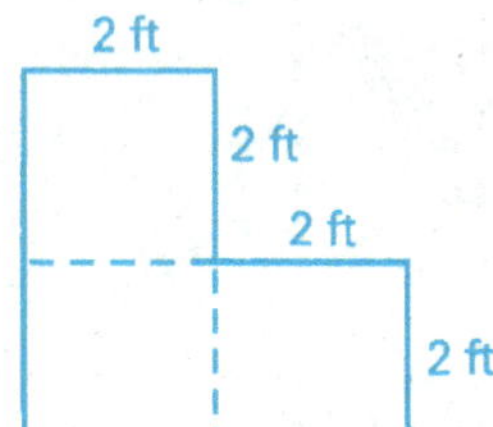

   The perimeter is _______ feet.

   The area is _______ square feet.

5. Use the front cover of this book. Measure its length and its width to the nearest inch. Find the perimeter of the cover. Find the area of the cover.

   The length of the cover is _______ inches.

   The width of the cover is _______ inches.

   The perimeter of the cover is _______ inches.

   The area of the cover is _______ square inches.

1.	2.
3.	4.
5.	

NAME ____________

Lesson 4 Volume

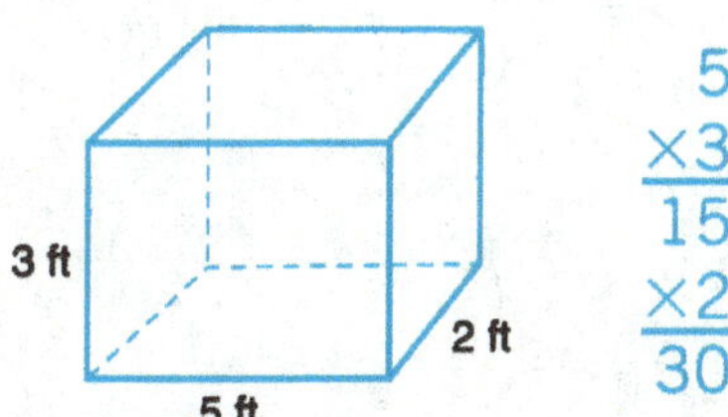

5
×3
15
×2
30

volume: 30 cubic feet

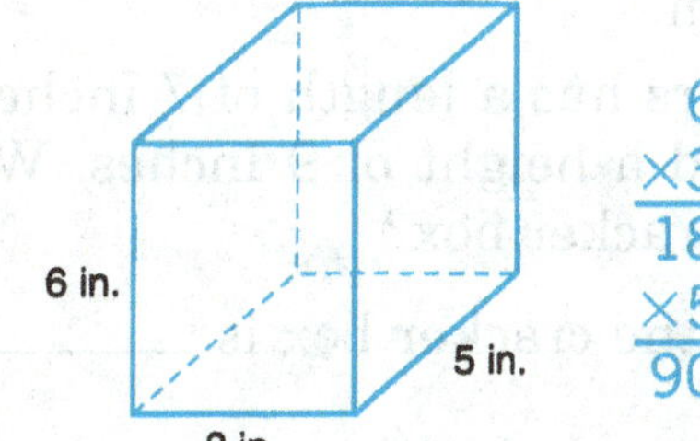

6
×3
18
×5
90

volume: 90 cubic inches

Find the volume of each rectangular solid.

a | *b* | *c*

1.

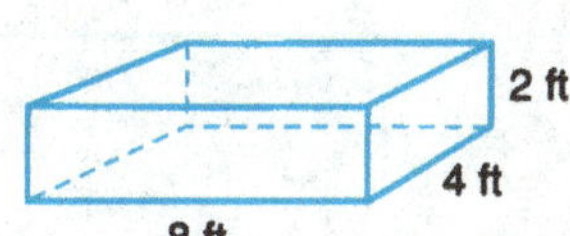

______ cubic feet

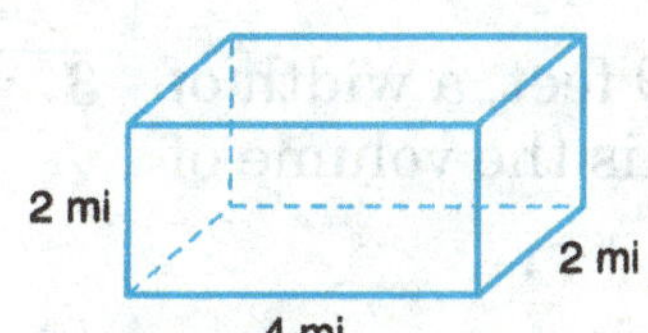

______ cubic miles

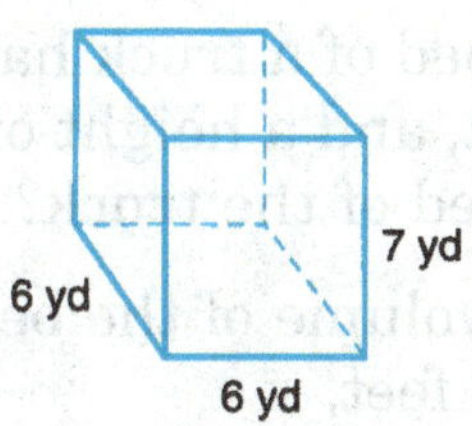

______ cubic yards

2.

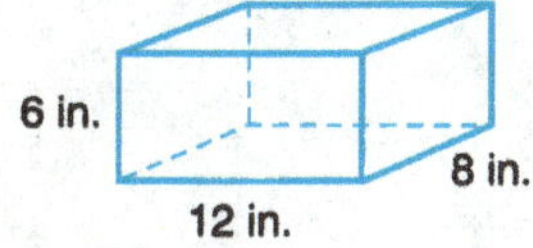

______ cubic inches

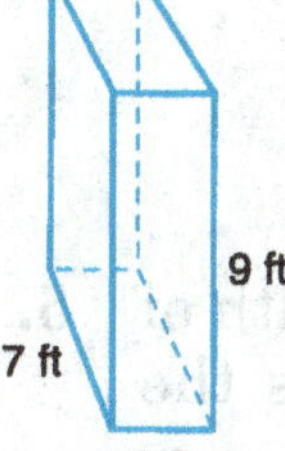

______ cubic feet

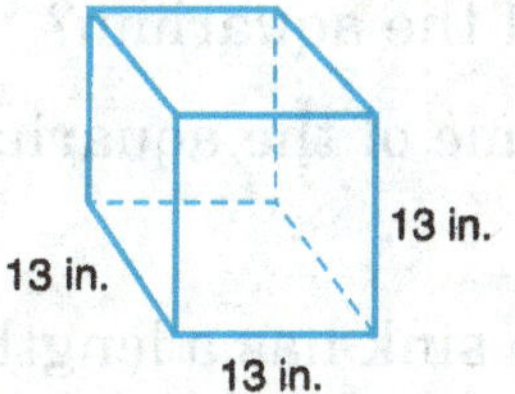

______ cubic inches

	length	*width*	*height*	*volume*
3.	4 mi	8 mi	3 mi	______ cubic miles
4.	7 in.	9 in.	11 in.	______ cubic inches
5.	8 yd	15 yd	5 yd	______ cubic yards
6.	18 ft	12 ft	10 ft	______ cubic feet
7.	10 in.	15 in.	6 in.	______ cubic inches

CHAPTER 9

Lesson 4 Problem Solving

Solve each problem.

1. A box of crackers has a length of 7 inches, a width of 3 inches, and a height of 9 inches. What is the volume of the cracker box?

The volume of the cracker box is _______ cubic inches.

1.

2. A swimming pool has a length of 5 yards, a width of 3 yards, and a depth of 1 yard. What is the volume of the swimming pool?

The volume of the swimming pool is _______ cubic yards.

2.

3. The bed of a truck has a length of 9 feet, a width of 5 feet, and a height of 3 feet. What is the volume of the bed of the truck?

The volume of the bed of the truck is _______ cubic feet.

3.

4. A fish aquarium has a length of 25 inches, a width of 12 inches, and a height of 11 inches. What is the volume of the aquarium?

The volume of the aquarium is _______ cubic inches.

4.

5. A kitchen sink has a length of 15 inches, a width of 14 inches, and a height of 9 inches. What is the volume of the kitchen sink?

The volume of the kitchen sink is _______ cubic inches.

5.

6. Find a cereal box that is a rectangular solid. Measure its length, width, and height. Find the volume of the cereal box.

length: _______________

width: _______________

height: _______________

volume: _______________

6.

NAME ____________________

Lesson 5 Capacity

1 pint (pt) = 2 cups
1 quart (qt) = 2 pt

1 gallon (gal) = 4 qt

6 pt = __?__ cups
1 pt = 2 cups
6 pt = (6 × 2) cups

6 pt = __12__ cups

2 gal = __?__ qt
1 gal = 4 qt
2 gal = (2 × 4) qt

2 gal = ______ qt

Complete the following.

	a	*b*
1.	8 cups = ______ pt	5 pt = ______ cups
2.	8 qt = ______ gal	16 pt = ______ qt
3.	16 qt = ______ pt	3 qt 1 pt = ______ pt
4.	5 qt 1 pt = ______ pt	6 gal 3 qt = ______ qt
5.	15 pt = ______ cups	7 pt 1 cup = ______ cups

6. Alex bought 6 pints of milk. He is going to give 1 cup of milk to each person. How many people can he serve?

He can serve ______ people.

6.

7. Mindy bought 6 pints of fruit juice. Sara bought 1 gallon and 1 quart of fruit juice. How many total quarts of fruit juice did each person buy? Who bought more? How many quarts more?

Mindy bought ______ quarts.

Sara bought ______ quarts.

______ bought ______ quarts more.

7.

CHAPTER 9

Lesson 5 Problem Solving

Solve each problem.

1. A fruit-drink recipe calls for 16 cups of water. How many pints of water is this? How many quarts?

   It is ______ pints of water.

   It is ______ quarts of water.

2. Ross counted 7 gallons of milk and 3 quarts of milk in the cooler. How many total quarts of milk was this? How many total pints of milk was this?

   It was ______ quarts of milk.

   It was ______ pints of milk.

3. Bri has 12 quarts and 1 pint of fruit drink. How many people can she serve at 1 pint per person? How many can she serve at 1 cup per person?

   She can serve ______ people at 1 pint each.

   She can serve ______ people at 1 cup each.

4. Gloria filled her aquarium with 8 gallons of water. How many quarts was that?

   It was ______ quarts of water.

5. It took 15 pints of hot tea to fill thirty china teacups. How many cups of hot tea is that?

   It is ______ cups of hot tea.

6. Jamal poured 3 gallons of water and 2 pints of cleaner into a bucket. How many quarts is that?

   It is ______ quarts of water and cleaner.

7. Rege needs 11 quarts of milk to make hot cocoa for his friends. How many gallons of milk should he buy?

   Rege should buy ______ gallons of milk.

1.
2.
3.
4.
5.
6.
7.

Lesson 6 Weight

1 pound (lb) = 16 ounces (oz)

3 lb 4 oz = ___?___ oz	144 oz = ___?___ lb
1 lb = 16 oz	16 oz = 1 lb
3 lb = (3 × 16) oz	144 oz = (144 ÷ 16) lb
3 lb 4 oz = (48 + 4) oz	144 oz = ______ lb
3 lb 4 oz = ___52___ oz	

Complete the following.

	a	*b*
1.	5 lb = ______ oz	48 oz = ______ lb
2.	96 oz = ______ lb	1 lb 6 oz = ______ oz
3.	6 lb = ______ oz	6 lb 2 oz = ______ oz
4.	112 oz = ______ lb	5 lb 6 oz = ______ oz
5.	240 oz = ______ lb	11 lb 2 oz = ______ oz

6. Anna's cat weighs 9 pounds. The cat's collar weighs 3 ounces. How many ounces does the cat weigh when it is wearing its collar?

The cat weighs ______ ounces with its collar.

6.

7. Juyong took a 5-pound stack of letters to the post office. Each of the 20 letters had the same weight. How many ounces did each letter weigh?

Each letter weighed ______ ounces.

7.

CHAPTER 9

Lesson 6 Problem Solving

Solve each problem.

1. Lauren and Juan have 3 pounds and 12 ounces of hamburger. How many ounces do they have?

   They have ______ ounces.

2. How many 4-ounce hamburgers can be made from the meat in problem **1**?

   ______ 4-ounce hamburgers can be made.

3. How many 3-ounce hamburgers can be made from the meat in problem **1**?

   ______ 3-ounce hamburgers can be made.

4. How many 6-ounce hamburgers can be made from the meat in problem **1**?

   ______ 6-ounce hamburgers can be made.

5. The candy shop sells 1-ounce squares of fudge. Teresa buys 1 pound of fudge. How many squares of fudge does Teresa buy?

   Teresa buys ______ squares of fudge.

6. Paul has 24 marbles in a bag. Each marble weighs 2 ounces. If he adds two more marbles to his bag, how much will the bag weigh?

   The bag will weigh ______ pounds ______ ounces.

7. Ramon wants to carry a backpack when he hikes. He has packed a 1-pound bag of granola, a map that weighs 12 ounces, and some apples that weigh 1 pound 2 ounces. How many ounces of these supplies has Ramon packed in his backpack?

   Ramon has packed ______ ounces in his backpack.

1.

2.

3.

4.

5.

6.

7.

NAME ____________

CHAPTER 9 PRACTICE TEST
Customary Measurement

Complete the following.

	a	*b*
1.	7 qt = ______ pt	9 ft = ______ in.
2.	18 cups = ______ pt	36 ft = ______ yd
3.	12 qt = ______ gal	10 yd = ______ in.
4.	5 gal 2 qt = ______ qt	5 qt 1 pt = ______ pt
5.	7 pt 1 cup = ______ cups	6 gal 3 qt = ______ qt

Find the perimeter of each figure below.

6.

a

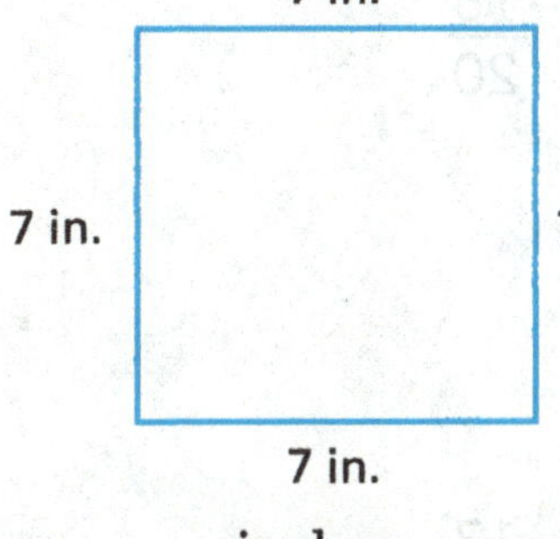

______ inches

b

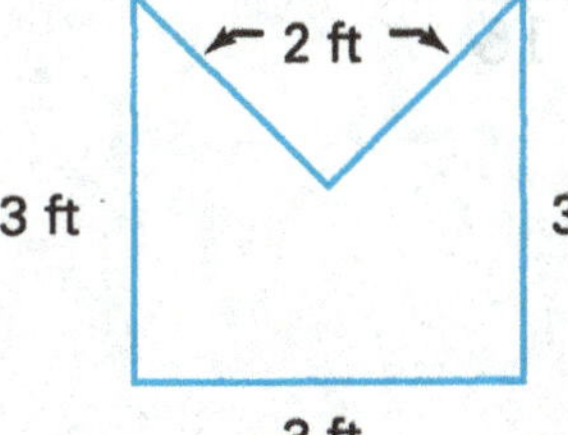

______ feet

c

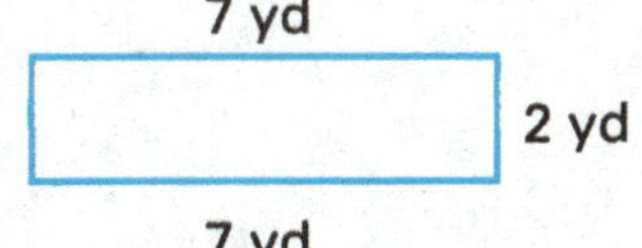

______ yards

Find the area of each rectangle below.

7.

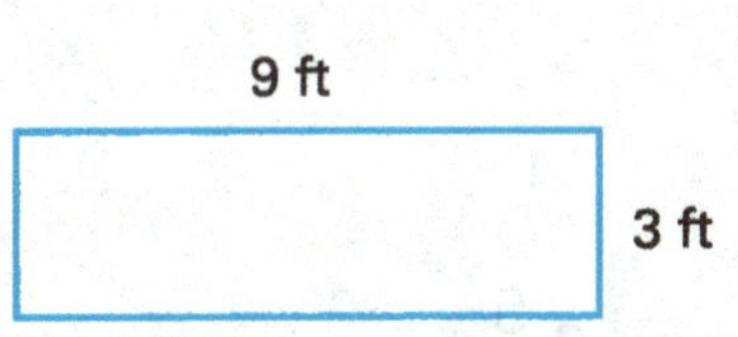

______ square feet

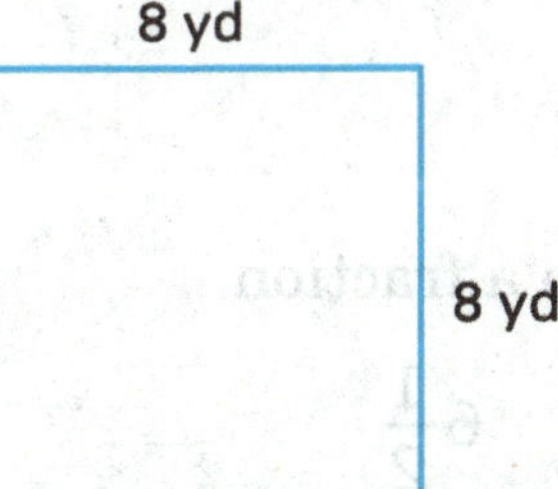

______ square yards

______ square inches

Find the volume.

	length	*width*	*height*	*volume*
8.	16 in.	4 in.	9 in.	______ sq in.
9.	11 ft	26 ft	3 ft	______ sq ft

CHAPTER 10 PRETEST
Fractions

Write the fraction that tells how much of each figure is colored.

	a	*b*	*c*	*d*
1.	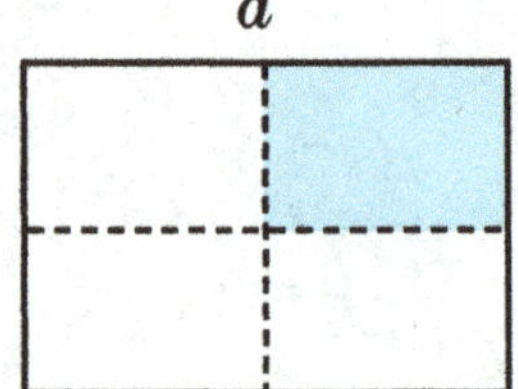	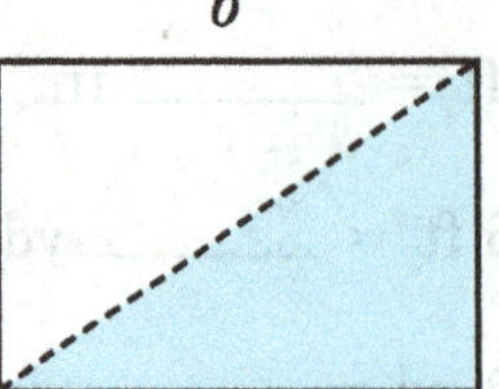	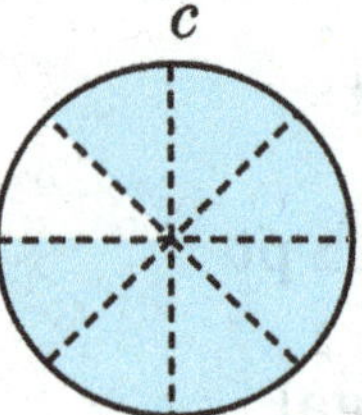	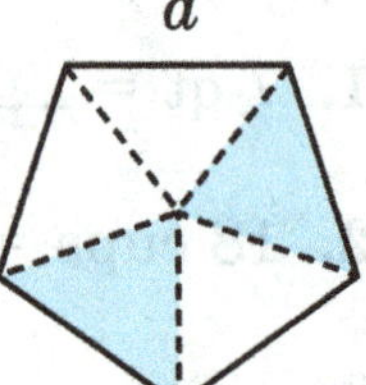
	______	______	______	______

Change each fraction to simplest form.

	a	*b*	*c*
2.	$\frac{4}{6}$	$\frac{8}{16}$	$\frac{15}{20}$

Rename as mixed numerals.

3.	$\frac{7}{6}$	$\frac{8}{3}$	$\frac{17}{5}$

Change each mixed numeral to a fraction.

4.	$3\frac{1}{4}$	$6\frac{1}{2}$	$3\frac{5}{6}$

Change each of the following to simplest form.

5.	$1\frac{6}{8}$	$\frac{10}{3}$	$4\frac{5}{2}$

NAME ______________________

Lesson 1 Writing Fractions

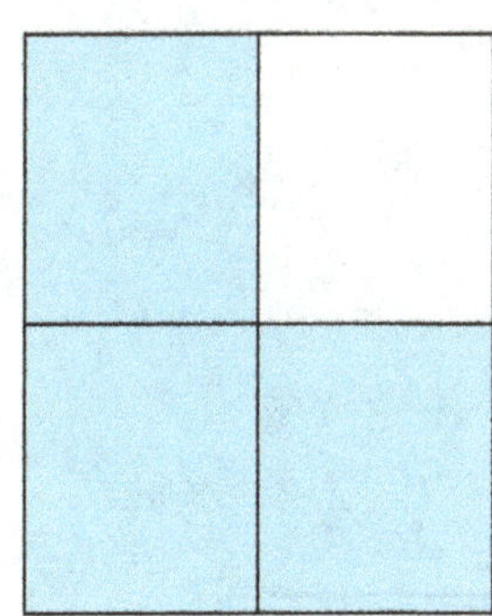

The figure is separated into 4 parts. Each part is the same size. 3 of the 4 parts are blue.
$\frac{3}{4}$ (read *three-fourths*) of the figure is blue.

____ of the 4 parts is not colored.

____ of the figure is not colored.

$\frac{3}{4}$ and $\frac{1}{4}$ are **fractions.**

On the first ____ beneath each figure, write the fraction that tells how much of the figure is blue. On the second ____, write the fraction that tells how much of the figure is not colored.

a *b* *c* *d*

1. 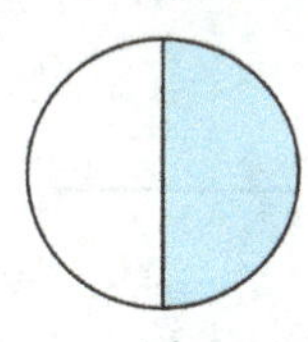 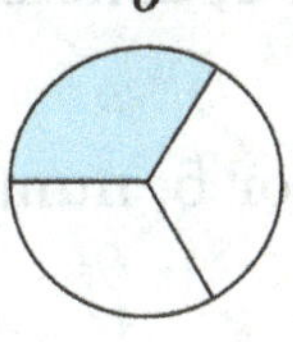 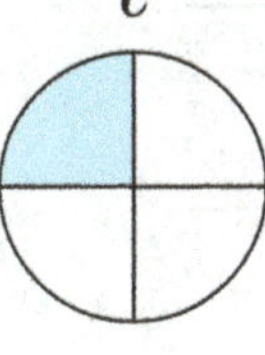 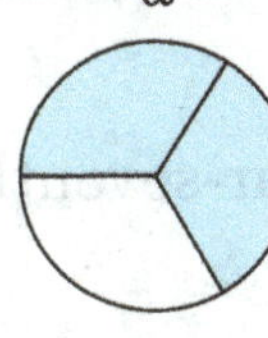

____ ____ ____ ____

____ ____ ____ ____

2. 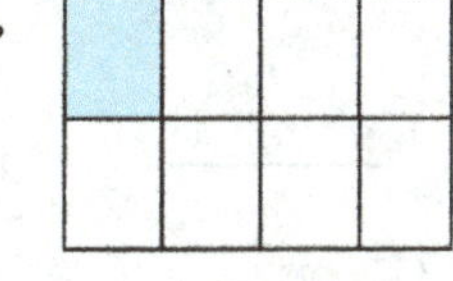 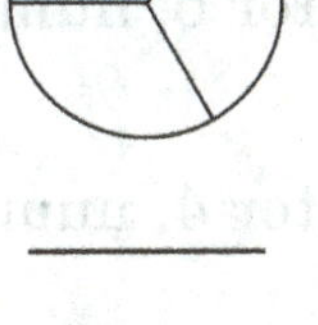 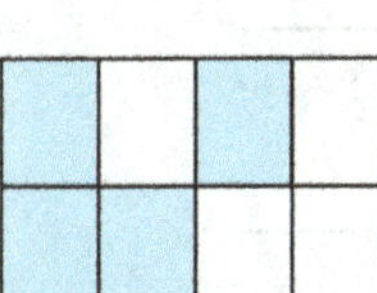 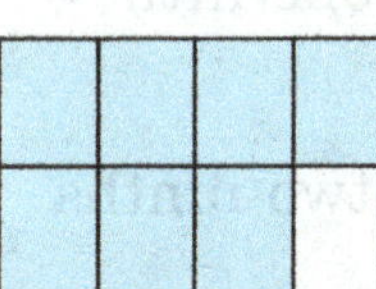

____ ____ ____ ____

____ ____ ____ ____

3. 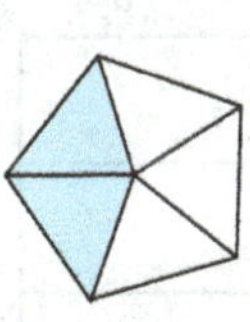 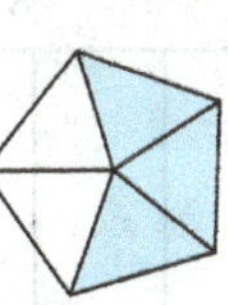 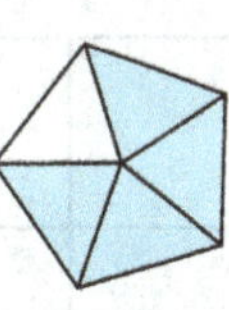 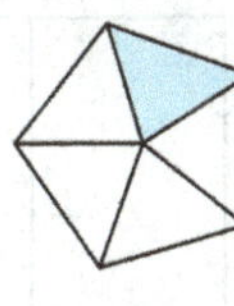

____ ____ ____ ____

____ ____ ____ ____

4. 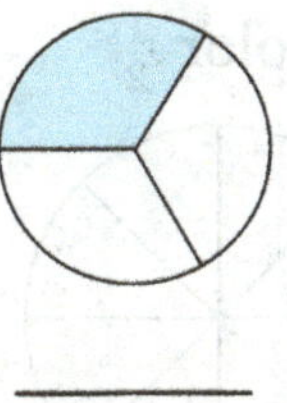 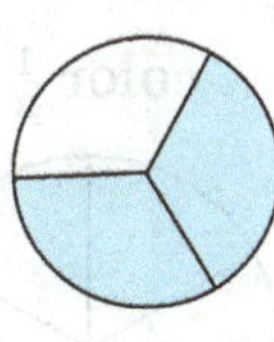 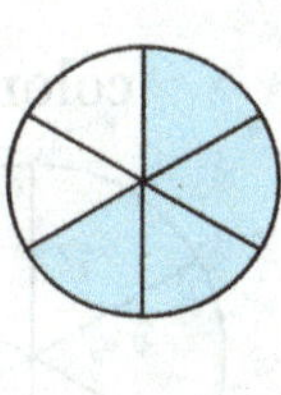

____ ____ ____ ____

____ ____ ____ ____

CHAPTER 10

NAME ______________________

Lesson 2 Writing Fractions

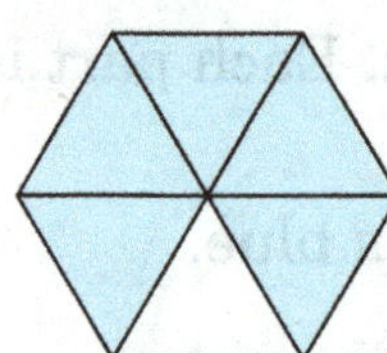

$\frac{5}{6}$ of the figure is colored.

$\frac{5}{6}$ ← numerator (5), ← denominator (6)

$\frac{1}{6}$ of the figure is not colored.

The denominator of $\frac{1}{6}$ is ______. The numerator of $\frac{1}{6}$ is ______.

Write a fraction for each of the following.

	a		*b*	
1.	three-fifths	______	numerator 2, denominator 3	______
2.	four-sevenths	______	denominator 5, numerator 4	______
3.	five-eighths	______	denominator 4, numerator 3	______
4.	one-fifth	______	numerator 1, denominator 6	______
5.	two-ninths	______	denominator 9, numerator 5	______

Color each figure as directed.

6.

a color $\frac{1}{2}$

b color $\frac{1}{4}$

c color $\frac{2}{3}$

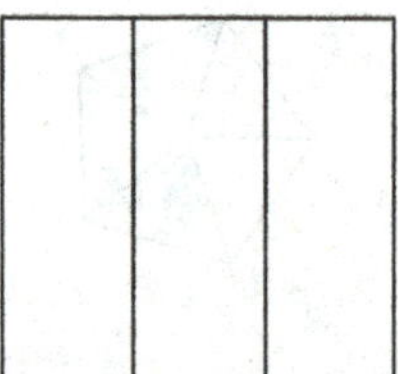

d color $\frac{1}{3}$

7.

color $\frac{2}{6}$

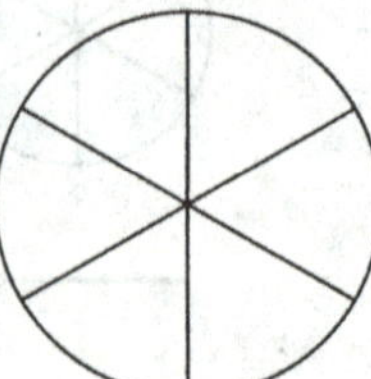

color $\frac{1}{3}$

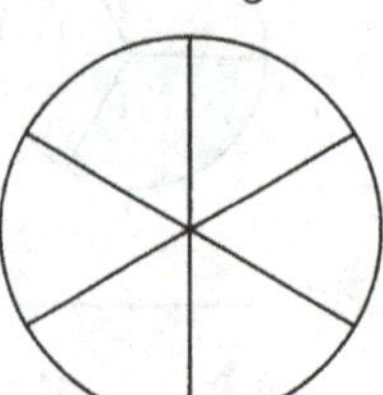

color $\frac{4}{8}$

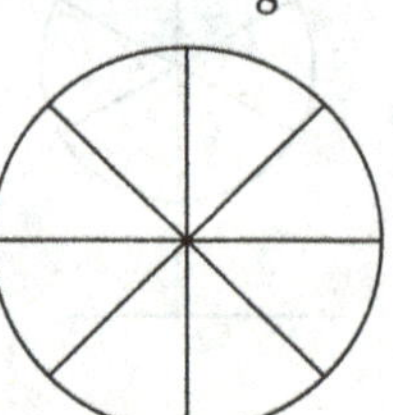

color $\frac{1}{2}$

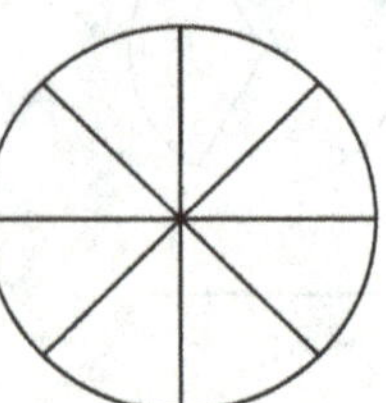

Lesson 3 Prime and Composite

A **factor** is a number that divides evenly (no remainder) into a given number.

A **prime number** is a number greater than 1 that has only 1 and itself as factors.

For example, 3 is a prime number because 1 and 3 are its only factors.

A **composite number** is a number that has more than two factors.

For example, 10 is a composite number because 1, 2, 5, and 10 are its factors.

List the factors of each number.

	a		*b*	
1.	5	____________	8	____________
2.	12	____________	11	____________
3.	6	____________	22	____________
4.	7	____________	13	____________
5.	15	____________	24	____________
6.	20	____________	9	____________

Identify each number as prime or composite.

	a		*b*	
7.	8	____________	17	____________
8.	14	____________	9	____________
9.	7	____________	11	____________
10.	12	____________	33	____________
11.	31	____________	27	____________
12.	25	____________	19	____________

CHAPTER 10

NAME ______________________

Lesson 4 Greatest Common Factor

The **greatest common factor** of two or more numbers is the largest factor they have in common.

What is the greatest common factor of 12 and 18?

List all the factors of both numbers.

12: (1), (2), (3), 4, (6), 12

18: (1), (2), (3), (6), 9, 18

Circle all their common factors.

The greatest common factor of 12 and 18 is 6.

List the common factors of each set of numbers.

	a	*b*
1.	8 and 14 ______________	15 and 30 ______________
2.	9 and 21 ______________	16 and 36 ______________

Find the greatest common factor of each set of numbers.

3.	15 and 20 ______	16 and 24 ______
4.	21 and 27 ______	20 and 28 ______
5.	36 and 48 ______	40 and 56 ______
6.	12, 24, and 36 ______	18, 30, and 42 ______

NAME ______________________

Lesson 5 Fractions in Simplest Form

A fraction is in simplest form when the only whole number that will divide the numerator and the denominator is 1.

$\frac{12}{18} = \frac{12 \div 6}{18 \div 6}$ ← Divide both the numerator and the denominator by the same number. → $\frac{12}{18} = \frac{12 \div 2}{18 \div 2}$

$= \frac{2}{3}$

$= \frac{6}{9}$ ← This fraction is not in simplest form, so continue dividing the numerator and the denominator until the fraction is in simplest form.

$= \frac{6 \div 3}{9 \div 3}$

$= \frac{2}{3}$

Change each fraction to simplest form.

	a	*b*	*c*
1.	$\frac{4}{6}$	$\frac{4}{16}$	$\frac{12}{15}$
2.	$\frac{12}{32}$	$\frac{8}{10}$	$\frac{15}{20}$
3.	$\frac{14}{16}$	$\frac{6}{8}$	$\frac{10}{16}$
4.	$\frac{6}{10}$	$\frac{3}{24}$	$\frac{8}{16}$
5.	$\frac{14}{21}$	$\frac{10}{12}$	$\frac{12}{16}$

CHAPTER 10

Lesson 5 Fractions in Simplest Form

Change each fraction to simplest form.

	a	*b*	*c*
1.	$\frac{4}{8}$	$\frac{3}{6}$	$\frac{2}{4}$
2.	$\frac{5}{10}$	$\frac{3}{15}$	$\frac{4}{20}$
3.	$\frac{4}{24}$	$\frac{8}{12}$	$\frac{6}{9}$
4.	$\frac{6}{21}$	$\frac{10}{25}$	$\frac{4}{12}$
5.	$\frac{12}{30}$	$\frac{12}{28}$	$\frac{16}{20}$
6.	$\frac{20}{24}$	$\frac{20}{36}$	$\frac{42}{49}$
7.	$\frac{21}{35}$	$\frac{15}{18}$	$\frac{24}{30}$
8.	$\frac{16}{24}$	$\frac{15}{35}$	$\frac{24}{32}$

NAME ______________________

Lesson 6 Improper Fractions

$\frac{17}{5}$ means $17 \div 5$ or $5\overline{)17}$.

$$\begin{array}{r} 3\frac{2}{5} \\ 5\overline{)17} \\ \underline{15} \\ 2 \end{array} \rightarrow 2 \div 5 = \frac{2}{5} \qquad \frac{17}{5} = 3\frac{2}{5}$$

$3\frac{2}{5}$ is a **mixed numeral.** It means $3 + \frac{2}{5}$.

Rename as mixed numerals.

	a	*b*	*c*
1.	$\frac{9}{4}$	$\frac{6}{5}$	$\frac{9}{8}$
2.	$\frac{8}{3}$	$\frac{9}{5}$	$\frac{7}{3}$
3.	$\frac{7}{4}$	$\frac{29}{6}$	$\frac{14}{3}$
4.	$\frac{15}{7}$	$\frac{12}{5}$	$\frac{19}{9}$
5.	$\frac{22}{7}$	$\frac{19}{2}$	$\frac{27}{5}$
6.	$\frac{35}{8}$	$\frac{43}{7}$	$\frac{55}{6}$

CHAPTER 10

NAME ______________________

Lesson 7 Renaming Numbers

Study how to change a mixed numeral to a fraction.

$$2\frac{1}{4} = \frac{(4 \times 2) + 1}{4} = \frac{8+1}{4} = \frac{9}{4}$$

Multiply the whole number by the denominator and add the numerator. Use the same denominator.

$$4\frac{2}{3} = \frac{(3 \times 4) + 2}{3} = \frac{12+2}{3} = \frac{14}{3}$$

Change each mixed numeral to a fraction.

	a	*b*	*c*
1.	$2\frac{1}{3}$	$3\frac{1}{2}$	$4\frac{3}{4}$
2.	$6\frac{4}{5}$	$3\frac{3}{8}$	$2\frac{5}{9}$
3.	$2\frac{1}{5}$	$1\frac{2}{7}$	$5\frac{3}{7}$
4.	$6\frac{5}{12}$	$7\frac{3}{10}$	$8\frac{6}{15}$

NAME ____________________

Lesson 8 Mixed Numerals

A mixed numeral is in simplest form when the fraction is in simplest form and names a number less than 1.

$$5\frac{4}{8} = 5 + \frac{4}{8}$$
$$= 5 + \frac{4 \div 4}{8 \div 4}$$
$$= 5 + \frac{1}{2}$$
$$= 5\frac{1}{2}$$

$$1\frac{18}{8} = 1 + \frac{18}{8}$$
$$= 1 + \frac{18 \div 2}{8 \div 2}$$
$$= 1 + \frac{9}{4}$$
$$\frac{9}{4} = 9 \div 4 = 2\frac{1}{4}$$
$$= 1 + 2\frac{1}{4}$$
$$= 3\frac{1}{4}$$

Change each mixed numeral to simplest form.

	a	*b*	*c*
1.	$3\frac{4}{6}$	$1\frac{4}{8}$	$2\frac{6}{8}$
2.	$4\frac{3}{12}$	$2\frac{6}{16}$	$1\frac{10}{12}$
3.	$1\frac{7}{5}$	$3\frac{9}{6}$	$2\frac{8}{6}$
4.	$1\frac{12}{10}$	$2\frac{15}{10}$	$4\frac{14}{6}$

CHAPTER 10

NAME ______________________

Lesson 9 Simplest Form

Change each fraction to simplest form.

	a	*b*	*c*
1.	$\frac{6}{14}$	$\frac{12}{27}$	$\frac{15}{25}$
2.	$\frac{4}{12}$	$\frac{28}{32}$	$\frac{15}{21}$

Change each of the following to a mixed numeral in simplest form.

3.	$\frac{9}{5}$	$\frac{8}{3}$	$\frac{12}{7}$
4.	$\frac{12}{8}$	$\frac{16}{6}$	$\frac{25}{15}$
5.	$1\frac{8}{10}$	$2\frac{7}{21}$	$3\frac{9}{15}$
6.	$4\frac{12}{14}$	$5\frac{8}{12}$	$2\frac{12}{16}$

NAME ______________________

CHAPTER 10 PRACTICE TEST
Fractions

Change each fraction to simplest form.

	a	*b*	*c*	*d*
1.	$\frac{4}{8}$	$\frac{5}{10}$	$\frac{6}{9}$	$\frac{3}{6}$
2.	$\frac{10}{15}$	$\frac{6}{8}$	$\frac{12}{18}$	$\frac{9}{12}$

Rename as mixed numerals.

3.	$\frac{5}{2}$	$\frac{7}{5}$	$\frac{9}{4}$	$\frac{16}{3}$

Change each mixed numeral to a fraction.

4.	$1\frac{1}{2}$	$1\frac{7}{8}$	$4\frac{2}{3}$	$5\frac{5}{6}$

Change each of the following to simplest form.

5.	$1\frac{8}{10}$	$\frac{18}{8}$	$1\frac{7}{3}$	$5\frac{12}{8}$

CHAPTER 10

NAME ____________________

CHAPTER 11 PRETEST
Multiplication of Fractions

Write each answer in simplest form.

	a	*b*	*c*
1.	$\frac{3}{7} \times \frac{2}{5}$	$\frac{3}{4} \times \frac{7}{8}$	$\frac{4}{5} \times \frac{4}{5}$
2.	$\frac{2}{3} \times \frac{7}{8}$	$\frac{5}{9} \times \frac{3}{5}$	$\frac{9}{10} \times \frac{5}{12}$
3.	$4 \times \frac{2}{3}$	$3 \times \frac{5}{6}$	$\frac{5}{8} \times 10$
4.	$3\frac{1}{5} \times 4$	$2\frac{1}{4} \times 8$	$6 \times 1\frac{5}{6}$
5.	$2\frac{1}{2} \times 2\frac{1}{3}$	$2\frac{1}{4} \times 1\frac{1}{5}$	$1\frac{1}{8} \times 3\frac{1}{3}$

CHAPTER 11

NAME ______________________

Lesson 1 Multiplication (using diagrams)

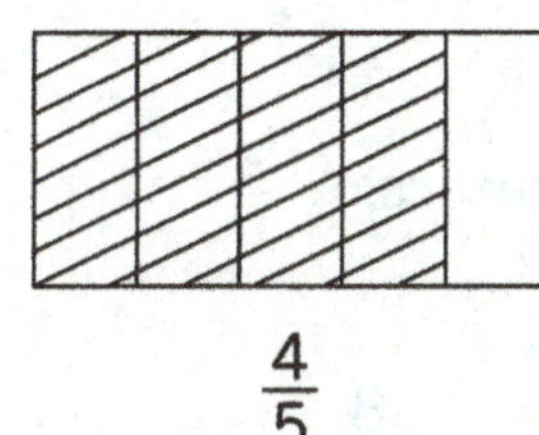

$\frac{4}{5}$

$\frac{2}{3}$ of $\frac{4}{5}$

5 parts in all. | 15 parts in all.

4 parts marked | ___ parts marked 

$\frac{4}{5}$ of the figure marked | ___ of the figure marked

$\frac{2}{3}$ of $\frac{4}{5} = \frac{8}{15}$

Complete the following.

	a		*b*	
1.	$\frac{1}{4}$	$\frac{1}{2}$ of $\frac{1}{4}$	$\frac{1}{2}$	$\frac{1}{2}$ of $\frac{1}{2}$

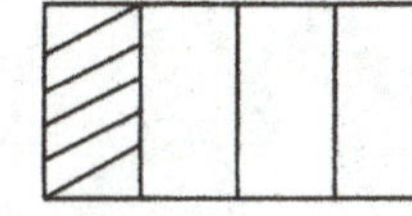 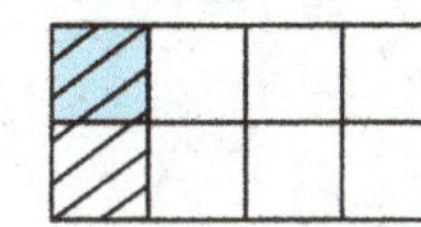 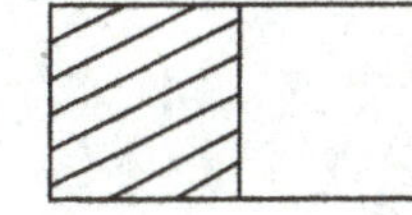 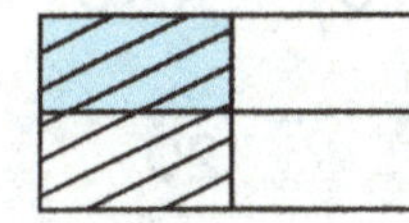

$\frac{1}{2}$ of $\frac{1}{4}$ = ______ $\frac{1}{2}$ of $\frac{1}{2}$ = ______

2. $\frac{1}{2}$ $\frac{1}{3}$ of $\frac{1}{2}$ $\frac{1}{3}$ $\frac{1}{2}$ of $\frac{1}{3}$

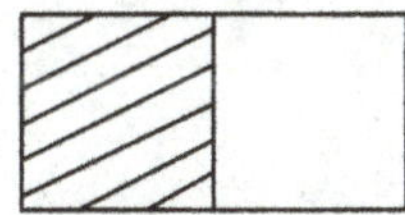 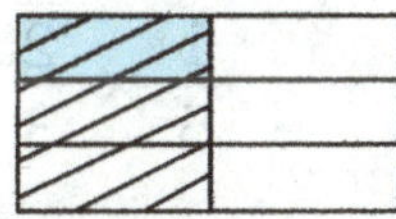 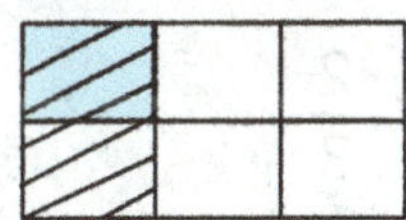

$\frac{1}{3}$ of $\frac{1}{2}$ = ______ $\frac{1}{2}$ of $\frac{1}{3}$ = ______

3. $\frac{1}{5}$ $\frac{1}{2}$ of $\frac{1}{5}$ $\frac{3}{5}$ $\frac{1}{2}$ of $\frac{3}{5}$

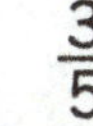

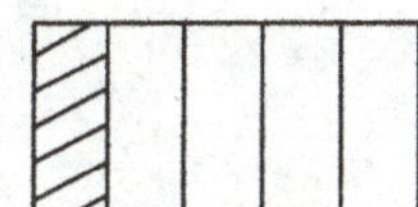 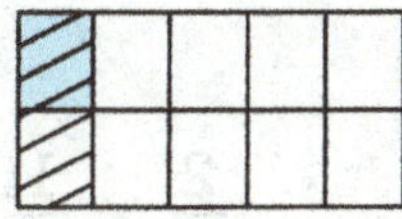 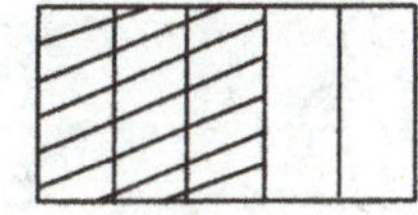 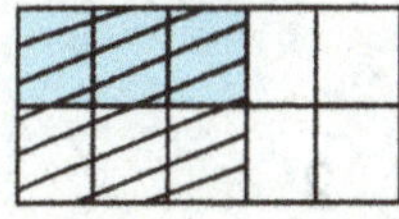

$\frac{1}{2}$ of $\frac{1}{5}$ = ______ $\frac{1}{2}$ of $\frac{3}{5}$ = ______

CHAPTER 11

NAME ____________________

Lesson 2 Multiplication

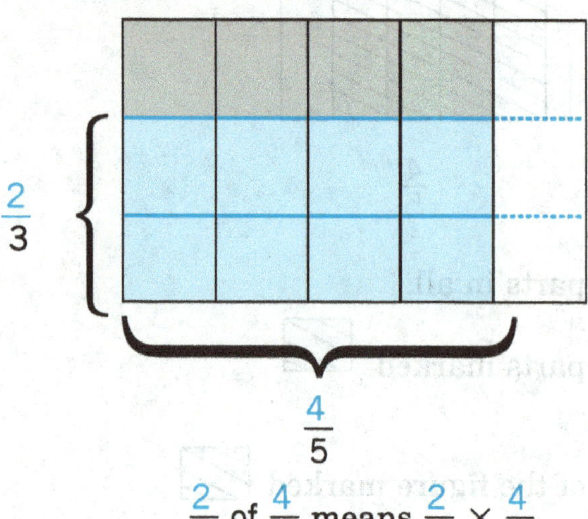

$\frac{2}{3}$ of $\frac{4}{5}$ means $\frac{2}{3} \times \frac{4}{5}$.

Multiply numerators.

$$\frac{2}{3} \times \frac{4}{5} = \frac{2 \times 4}{3 \times 5} = \frac{8}{15}$$

Multiply denominators.

Multiply as shown.

	a	*b*	*c*
1.	$\frac{1}{4} \times \frac{3}{5} = \frac{1 \times 3}{4 \times 5} = \frac{3}{20}$	$\frac{2}{3} \times \frac{1}{5}$	$\frac{1}{6} \times \frac{5}{8}$
2.	$\frac{3}{7} \times \frac{1}{4}$	$\frac{5}{9} \times \frac{1}{2}$	$\frac{6}{7} \times \frac{2}{5}$
3.	$\frac{4}{5} \times \frac{2}{3}$	$\frac{7}{8} \times \frac{1}{6}$	$\frac{1}{5} \times \frac{2}{3}$
4.	$\frac{2}{5} \times \frac{1}{7}$	$\frac{5}{6} \times \frac{1}{2}$	$\frac{2}{3} \times \frac{5}{7}$
5.	$\frac{2}{3} \times \frac{2}{5}$	$\frac{5}{8} \times \frac{3}{4}$	$\frac{2}{5} \times \frac{1}{3}$

Lesson 3 Multiplication

$$\frac{4}{5} \times \frac{1}{2} = \frac{4 \times 1}{5 \times 2}$$ ← Multiply the numerators. → $$\frac{3}{10} \times \frac{5}{6} = \frac{3 \times 5}{10 \times 6}$$
← Multiply the denominators. →

$$= \frac{4}{10} \qquad\qquad = \frac{15}{60}$$

$$= \frac{2}{5}$$ ← If necessary, change the answer to simplest form. → $$= \frac{1}{4}$$

Write each answer in simplest form.

	a	*b*	*c*
1.	$\frac{5}{7} \times \frac{1}{4}$	$\frac{3}{5} \times \frac{1}{2}$	$\frac{7}{8} \times \frac{3}{4}$
2.	$\frac{3}{7} \times \frac{2}{5}$	$\frac{1}{4} \times \frac{7}{8}$	$\frac{3}{5} \times \frac{4}{9}$
3.	$\frac{4}{7} \times \frac{3}{8}$	$\frac{9}{10} \times \frac{5}{6}$	$\frac{5}{9} \times \frac{6}{10}$
4.	$\frac{8}{15} \times \frac{5}{12}$	$\frac{5}{12} \times \frac{16}{25}$	$\frac{4}{9} \times \frac{9}{14}$
5.	$\frac{6}{7} \times \frac{2}{3}$	$\frac{7}{8} \times \frac{11}{12}$	$\frac{3}{10} \times \frac{7}{8}$

CHAPTER 11

Lesson 3 Problem Solving

Solve. Write each answer in simplest form.

1. Jeff had $\frac{3}{4}$ yard of string. He used $\frac{2}{3}$ of the string to tie a package. How much string did he use?

   He used ________ yard.

2. Julia lives $\frac{7}{8}$ mile from work. She has walked $\frac{4}{5}$ of the way to work. How far has she walked?

   She has walked ________ mile.

3. Marla bought $\frac{1}{2}$ gallon of milk. She drank $\frac{1}{4}$ of it. How much milk did she drink?

   She drank ________ gallon.

4. Hakeem bought $\frac{3}{4}$ pound of cheese. He ate $\frac{1}{3}$ of it. How much cheese did he eat?

   He ate ________ pound.

5. Five-sixths of a room is now painted. Carlos did $\frac{2}{5}$ of the painting. How much of the room did he paint?

   He painted ________ of the room.

6. The lawn is $\frac{1}{2}$ mowed. Melinda did $\frac{2}{3}$ of the mowing. How much of the lawn did she mow?

   She mowed ________ of the lawn.

7. A lawn mower uses $\frac{3}{4}$ gallon of fuel each hour. How much fuel will it use in $\frac{1}{2}$ hour?

   It will use ________ gallon.

1.
2.
3.
4.
5.
6.
7.

NAME ______________________

Lesson 4 Multiplication (by whole numbers)

$4 \times \frac{2}{5} = \frac{4}{1} \times \frac{2}{5}$ Name the whole number as a fraction. $\frac{5}{8} \times 6 = \frac{5}{8} \times \frac{6}{1}$

$= \frac{4 \times 2}{1 \times 5}$ Multiply the fractions. $= \frac{5 \times 6}{8 \times 1}$

$= \frac{8}{5}$ $= \frac{30}{8}$

$5\overline{)8}$ gives $1\frac{3}{5}$; $8\overline{)30}$ gives $3\frac{3}{4}$

$= 1\frac{3}{5}$ Change the answer to simplest form. $= 3\frac{3}{4}$

Write each answer in simplest form.

	a	*b*	*c*
1.	$5 \times \frac{3}{7}$	$9 \times \frac{7}{8}$	$7 \times \frac{5}{6}$
2.	$\frac{2}{3} \times 5$	$\frac{7}{8} \times 9$	$\frac{4}{5} \times 12$
3.	$8 \times \frac{3}{4}$	$9 \times \frac{5}{6}$	$4 \times \frac{4}{5}$
4.	$\frac{7}{8} \times 12$	$\frac{3}{5} \times 10$	$\frac{5}{6} \times 14$

CHAPTER 11

Lesson 4 Problem Solving

Solve. Write each answer in simplest form.

1. A boy weighs 60 pounds on Earth. He would weigh only $\frac{1}{6}$ of that on the moon. How much would he weigh on the moon?

   He would weigh ________ pounds.

   1.

2. A woman weighs 120 pounds on Earth. How much would she weigh on the moon?

   She would weigh ________ pounds.

   2.

3. A dog weighs 20 pounds on Earth. It would weigh only $\frac{2}{5}$ of that on Mars. How much would the dog weigh on Mars?

   It would weigh ________ pounds.

   3.

4. How much would the boy in problem **1** weigh on Mars?

   He would weigh ________ pounds.

   4.

5. How much would the woman in problem **2** weigh on Mars?

   She would weigh ________ pounds.

   5.

6. A rock weighs 10 pounds on Earth. It would weigh only $\frac{7}{8}$ of that on Venus. How much would it weigh on Venus?

   It would weigh ________ pounds.

   6.

7. How much would the dog in problem **3** weigh on Venus?

   It would weigh ________ pounds.

   7.

NAME ________________

Lesson 5 Multiplication (mixed numerals)

$2\frac{1}{6} \times 8 = \frac{13}{6} \times \frac{8}{1}$ — Change the mixed numeral to a fraction. Name the whole number as a fraction.

$= \frac{13 \times 8}{6 \times 1}$ — Multiply.

$= \frac{104}{6}$

$= 17\frac{1}{3}$ — Change the answer to simplest form.

Write each answer in simplest form.

	a	*b*	*c*
1.	$4\frac{1}{2} \times 5$	$1\frac{3}{4} \times 7$	$3 \times 2\frac{1}{8}$
2.	$2\frac{2}{3} \times 6$	$1\frac{7}{8} \times 6$	$4 \times 2\frac{3}{8}$
3.	$2\frac{4}{5} \times 7$	$10 \times 2\frac{4}{15}$	$8\frac{1}{7} \times 4$
4.	$8 \times 2\frac{5}{6}$	$3\frac{2}{7} \times 14$	$3\frac{1}{3} \times 7$

CHAPTER 11

Lesson 5 Problem Solving

Solve. Write each answer in simplest form.

1. Some square tiles measure $3\frac{1}{2}$ inch on each side. Seven tiles are placed in a row. How long is the row of tiles? The row would be ______ inches long.	1.
2. Suppose that ten tiles like those in problem **1** were placed in a row. How long would that row of tiles be? It would be ______ inches long.	2.
3. There are 5 boxes and each one weighs $1\frac{3}{4}$ pounds. How many pounds do all the boxes weigh? All the boxes weigh ______ pounds.	3.
4. Each board is $1\frac{5}{8}$ inches thick. Six boards are stacked on top of each other. How high is the stack? The stack of boards is ______ inches high.	4.
5. Suppose it takes $2\frac{5}{6}$ hours to make an orbit around the moon. How long would it take to make 9 orbits? It would take ______ hours.	5.
6. There are a dozen boxes of nails in each carton. Each box of nails weighs $2\frac{1}{2}$ pounds. How much would a carton of nails weigh? One carton would weigh ______ pounds.	6.
7. In problem **6,** suppose there are only six boxes left in the carton. How much would that carton weigh? It would weigh ______ pounds.	7.
8. Each straight piece of road-racing track is $5\frac{3}{8}$ inches long. What would be the total length of track if Jill lays ten pieces of straight track end-to-end? The total length would be ______ inches.	8.

NAME ______________________

Lesson 6 Multiplication (mixed numerals)

$1\frac{1}{2} \times 2\frac{1}{4} = \frac{3}{2} \times \frac{9}{4}$ — **Change both mixed numerals to fractions.**

$= \frac{3 \times 9}{2 \times 4}$ — **Multiply.**

$= \frac{27}{8}$

$= 3\frac{3}{8}$ — Change to simplest form.

Write each answer in simplest form.

	a	*b*	*c*
1.	$3\frac{1}{8} \times 1\frac{2}{3}$	$1\frac{1}{6} \times 2\frac{1}{2}$	$1\frac{4}{5} \times 1\frac{3}{4}$
2.	$2\frac{2}{3} \times 4\frac{1}{5}$	$2\frac{1}{2} \times 1\frac{1}{7}$	$1\frac{3}{5} \times 1\frac{1}{6}$
3.	$1\frac{3}{5} \times 3\frac{3}{4}$	$2\frac{1}{4} \times 3\frac{1}{3}$	$4\frac{1}{2} \times 2\frac{2}{3}$
4.	$2\frac{2}{5} \times 2\frac{1}{4}$	$1\frac{3}{8} \times 1\frac{3}{7}$	$2\frac{4}{5} \times 2\frac{6}{7}$

CHAPTER 11

Lesson 6 Problem Solving

Solve. Write each answer in simplest form.

1. A rectangle is $4\frac{1}{2}$ feet long and $1\frac{3}{4}$ feet wide. Find the area of the rectangle.

   The area is ________ square feet.

   1.

2. A rectangular picture is $1\frac{3}{4}$ inches long and $3\frac{1}{2}$ inches wide. Find the area of the picture.

   The area is ________ square inches.

   2.

3. A rectangular window is $2\frac{1}{2}$ feet long and $4\frac{1}{2}$ feet wide. Find the area of the window.

   The area is ________ square feet.

   3.

4. Each side of a square floor is $10\frac{1}{2}$ feet long. Find the area of that floor.

   The area is ________ square feet.

   4.

5. A boat was traveling $12\frac{1}{2}$ miles each hour. At that rate, how many miles would it travel in $1\frac{1}{2}$ hours?

   It would travel ________ miles.

   5.

6. How many miles would the boat in problem **5** travel in 4 hours?

   It would travel ________ miles.

   6.

7. How many miles would the boat in problem **5** travel in $5\frac{1}{4}$ hours?

   It would travel ________ miles.

   7.

Lesson 7 Multiplication Review

Write each answer in simplest form.

	a	*b*	*c*	*d*
1.	$\frac{3}{4} \times \frac{1}{5}$	$\frac{2}{7} \times \frac{3}{5}$	$\frac{2}{3} \times \frac{1}{5}$	$\frac{5}{12} \times \frac{7}{8}$
2.	$\frac{6}{7} \times \frac{1}{3}$	$\frac{4}{7} \times \frac{5}{6}$	$\frac{3}{8} \times \frac{2}{9}$	$\frac{3}{4} \times \frac{5}{12}$
3.	$6 \times \frac{2}{5}$	$\frac{2}{7} \times 4$	$8 \times \frac{3}{4}$	$\frac{3}{8} \times 6$
4.	$6\frac{2}{5} \times 5$	$6\frac{7}{8} \times 16$	$4 \times 5\frac{5}{6}$	$8 \times 2\frac{1}{12}$
5.	$3\frac{1}{8} \times 3\frac{1}{5}$	$4\frac{2}{3} \times 1\frac{4}{5}$	$2\frac{1}{2} \times 4\frac{2}{3}$	$1\frac{3}{5} \times 1\frac{1}{4}$

CHAPTER 11

Lesson 7 Problem Solving

Solve. Write each answer in simplest form.

1. Zoe spent $\frac{2}{3}$ hour doing homework. She spent $\frac{3}{4}$ of this time reading. How long did she spend reading?

   She spent ________ hour reading.

1.

2. A rectangular picture is $8\frac{1}{2}$ inches long and 10 inches wide. Find the area of the picture.

   The area is ________ square inches.

2.

3. In one hour a machine can produce $\frac{9}{10}$ pound of silver. Suppose the machine breaks down after $\frac{1}{3}$ hour. How many pounds of silver are processed?

   ________ pound of silver is processed.

3.

4. A certain book is $\frac{7}{8}$ inch thick. Ten of these books are placed on top of each other. How high is the stack?

   The stack of books is ________ inches high.

4.

5. A large box of Lotsa-clean detergent weighs $6\frac{3}{4}$ pounds. There are 12 of these boxes in a carton. How much would a carton weigh?

   A carton would weigh ________ pounds.

5.

6. There are $4\frac{1}{2}$ pounds of dog food in each bag. How many pounds of dog food would be in 3 bags?

   There would be ________ pounds in 3 bags.

6.

7. Chloe gained 3 pounds in six months. Matt gained $3\frac{1}{9}$ times as many pounds as Chloe. How many pounds did Matt gain?

   Matt gained ________ pounds.

7.

NAME ______________________

CHAPTER 11 PRACTICE TEST
Multiplication of Fractions

Write each answer in simplest form.

	a	*b*	*c*
1.	$\frac{7}{8} \times \frac{5}{6}$	$\frac{4}{5} \times \frac{3}{7}$	$\frac{2}{3} \times \frac{1}{5}$
2.	$\frac{2}{3} \times \frac{5}{6}$	$\frac{8}{9} \times \frac{3}{8}$	$\frac{2}{5} \times \frac{15}{16}$
3.	$8 \times \frac{3}{5}$	$9 \times \frac{5}{6}$	$\frac{3}{4} \times 20$
4.	$2\frac{2}{5} \times 4$	$4\frac{1}{4} \times 6$	$3 \times 1\frac{2}{9}$
5.	$\frac{2}{3} \times 1\frac{4}{5}$	$7\frac{1}{2} \times \frac{4}{5}$	$6\frac{1}{4} \times \frac{2}{5}$
6.	$1\frac{3}{5} \times 1\frac{1}{3}$	$2\frac{1}{2} \times 3\frac{1}{3}$	$2\frac{1}{6} \times 1\frac{1}{8}$

CHAPTER 11

CHAPTER 12 PRETEST
Addition of Fractions

Write each answer in simplest form.

	a	*b*	*c*	*d*
1.	$\frac{1}{6} + \frac{1}{6}$	$\frac{3}{8} + \frac{1}{8}$	$\frac{5}{9} + \frac{2}{9}$	$\frac{7}{12} + \frac{5}{12}$
2.	$\frac{5}{6} + \frac{1}{3}$	$\frac{7}{8} + \frac{1}{2}$	$\frac{7}{10} + \frac{2}{5}$	$\frac{3}{5} + \frac{1}{4}$
3.	$7\frac{1}{2} + 3\frac{1}{4}$	$6\frac{7}{10} + 1\frac{1}{5}$	$5\frac{1}{3} + \frac{3}{4}$	$4\frac{1}{3} + 2\frac{1}{2}$
4.	$1\frac{5}{8} + 4\frac{1}{6}$	$5\frac{3}{4} + \frac{1}{5}$	$\frac{7}{12} + \frac{5}{6}$	$\frac{1}{12} + 6\frac{3}{4}$
5.	$\frac{2}{3} + \frac{3}{4}$	$9\frac{3}{8} + \frac{1}{4}$	$3\frac{4}{5} + 1\frac{3}{10}$	$4\frac{2}{3} + 5\frac{5}{6}$

Lesson 1 Addition (fractions)

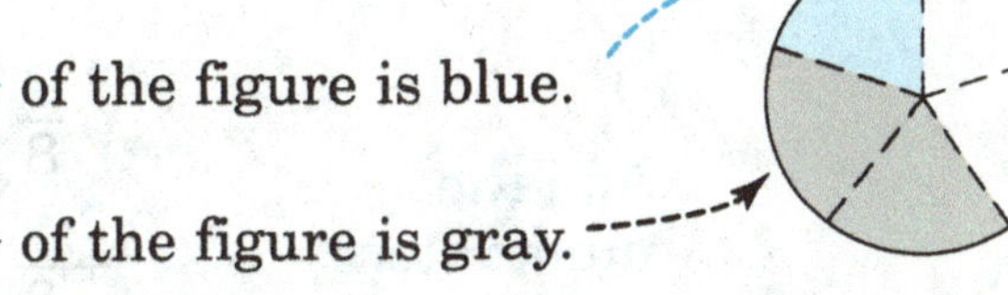

$\frac{1}{5}$ of the figure is blue.

$\frac{2}{5}$ of the figure is gray.

$\frac{3}{5}$ of the figure is colored. $\frac{1}{5} + \frac{2}{5} = \frac{3}{5}$

Complete the following.

	a	*b*	*c*
1.	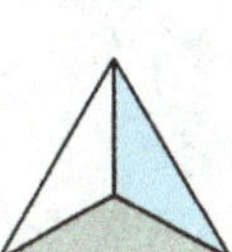 $\frac{1}{3} + \frac{1}{3} =$	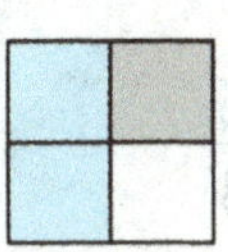 $\frac{2}{4} + \frac{1}{4} =$	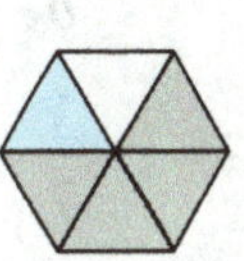 $\frac{1}{6} + \frac{4}{6} =$
2.	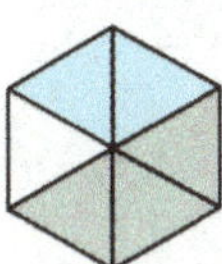 $\frac{2}{6} + \frac{3}{6} =$	$\frac{2}{7} + \frac{2}{7} =$	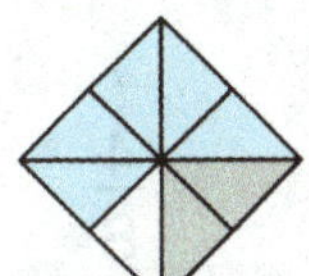 $\frac{5}{8} + \frac{2}{8} =$
3.	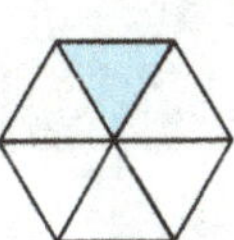 $\frac{1}{6} + \frac{0}{6} =$	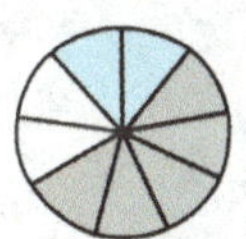 $\frac{2}{9} + \frac{5}{9} =$	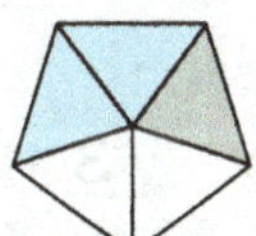 $\frac{2}{5} + \frac{1}{5} =$
4.	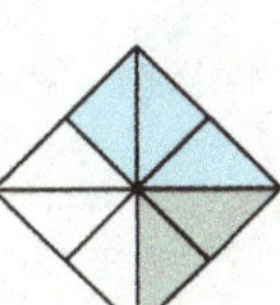 $\frac{3}{8} + \frac{2}{8} =$	$\frac{3}{5} + \frac{1}{5} =$	$\frac{4}{9} + \frac{4}{9} =$

CHAPTER 12

NAME ______________________

Lesson 2 Addition (like denominators)

Study how to add two fractions that have the same denominator.

Add the numerators.

$$\frac{3}{8} + \frac{2}{8} = \frac{3+2}{8} = \frac{5}{8}$$

Use the same denominator.

Add the numerators.

Use the same denominator.

Add.

	a	b	c	d	e
1.	$\frac{1}{3} + \frac{1}{3}$	$\frac{2}{7} + \frac{4}{7}$	$\frac{5}{8} + \frac{2}{8}$	$\frac{1}{4} + \frac{2}{4}$	$\frac{2}{5} + \frac{2}{5}$
2.	$\frac{4}{9} + \frac{3}{9}$	$\frac{4}{8} + \frac{1}{8}$	$\frac{1}{6} + \frac{4}{6}$	$\frac{3}{7} + \frac{3}{7}$	$\frac{2}{10} + \frac{5}{10}$
3.	$\frac{2}{5} + \frac{1}{5}$	$\frac{3}{6} + \frac{2}{6}$	$\frac{2}{8} + \frac{1}{8}$	$\frac{2}{7} + \frac{2}{7}$	$\frac{2}{9} + \frac{2}{9}$
4.	$\frac{1}{9} + \frac{4}{9}$	$\frac{1}{7} + \frac{4}{7}$	$\frac{6}{8} + \frac{1}{8}$	$\frac{1}{5} + \frac{1}{5}$	$\frac{3}{7} + \frac{1}{7}$

NAME ______________________

Lesson 3 Addition (like denominators)

$$\begin{array}{r} \frac{7}{10} \\ +\frac{9}{10} \\ \hline \frac{16}{10} \end{array} = 1\frac{3}{5}$$

Add.

Change to simplest form.

$$\begin{array}{r} \frac{1}{12} \\ +\frac{11}{12} \\ \hline \frac{12}{12} \end{array} = 1$$

Add. Write each answer in simplest form.

	a	*b*	*c*	*d*
1.	$\frac{2}{3} + \frac{2}{3}$	$\frac{4}{5} + \frac{3}{5}$	$\frac{2}{9} + \frac{1}{9}$	$\frac{1}{4} + \frac{1}{4}$
2.	$\frac{1}{8} + \frac{5}{8}$	$\frac{3}{10} + \frac{9}{10}$	$\frac{3}{4} + \frac{3}{4}$	$\frac{7}{12} + \frac{11}{12}$
3.	$\frac{1}{2} + \frac{1}{2}$	$\frac{6}{7} + \frac{5}{7}$	$\frac{7}{8} + \frac{7}{8}$	$\frac{5}{6} + \frac{1}{6}$
4.	$\frac{3}{5} + \frac{3}{5}$	$\frac{5}{12} + \frac{7}{12}$	$\frac{8}{9} + \frac{5}{9}$	$\frac{7}{10} + \frac{9}{10}$

CHAPTER 12

NAME ______________________

Lesson 4 Addition (mixed numerals)

$4\frac{5}{8}$	Add the fractions.	$6\frac{7}{10}$
$+2\frac{1}{8}$	Add the whole numbers.	$+2\frac{9}{10}$
$6\frac{6}{8} = 6\frac{3}{4}$	Change to simplest form.	$8\frac{16}{10} = 9\frac{3}{5}$

Add. Write each answer in simplest form.

	a	*b*	*c*	*d*
1.	$1\frac{2}{5} + 2\frac{1}{5}$	$4\frac{1}{6} + 2\frac{1}{6}$	$3\frac{1}{10} + 2\frac{3}{10}$	$19\frac{3}{8} + 7\frac{1}{8}$
2.	$5\frac{3}{4} + 1\frac{3}{4}$	$6\frac{2}{3} + 1\frac{1}{3}$	$2\frac{9}{10} + 1\frac{7}{10}$	$26\frac{4}{5} + 13\frac{3}{5}$
3.	$4\frac{1}{2} + 2\frac{1}{2}$	$3\frac{5}{6} + 4\frac{5}{6}$	$8\frac{7}{12} + 4\frac{11}{12}$	$36\frac{7}{8} + 27\frac{5}{8}$
4.	$7\frac{2}{3} + 6\frac{2}{3}$	$9\frac{2}{5} + 4\frac{4}{5}$	$11\frac{3}{10} + 6\frac{7}{10}$	$58\frac{7}{9} + 31\frac{5}{9}$

NAME ____________________

Lesson 5 Renaming Fractions

By separating the figure in different ways, you can write different fractions to tell how much is blue.

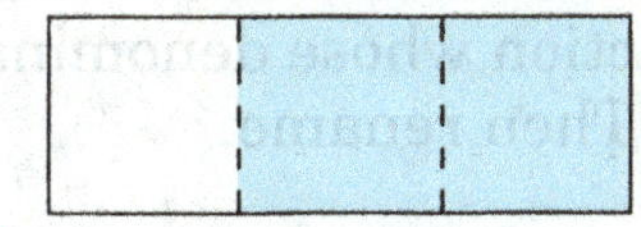

$\frac{2}{3}$ of the figure is blue.

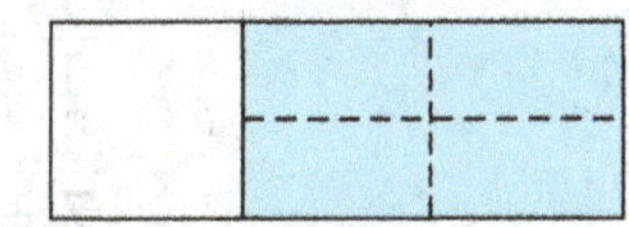

$\frac{4}{6}$ of the figure is blue.

$$\frac{2}{3} = \frac{4}{6}$$

$\frac{2}{3} = \frac{\square}{6}$

$\frac{2}{3} = \frac{2 \times 2}{3 \times 2}$

$\frac{2}{3} = \frac{4}{6}$

Choose 2 so the new denominator is 6.

Multiply the numerator and the denominator by the same number.

$\frac{2}{3} = \frac{\square}{9}$

$\frac{2}{3} = \frac{2 \times 3}{3 \times 3}$

$\frac{2}{3} = \frac{6}{9}$

Choose 3 so the new denominator is 9.

Rename.

	a	*b*	*c*
1.	$\frac{2}{3} = \frac{\square}{12}$	$\frac{3}{4} = \frac{\square}{8}$	$\frac{5}{6} = \frac{\square}{12}$
2.	$\frac{1}{2} = \frac{\square}{10}$	$\frac{2}{5} = \frac{\square}{10}$	$\frac{3}{5} = \frac{\square}{15}$
3.	$\frac{3}{4} = \frac{\square}{12}$	$\frac{3}{8} = \frac{\square}{16}$	$\frac{4}{5} = \frac{\square}{20}$

CHAPTER 12

Lesson 5 Renaming Fractions

$$\frac{7}{8} = \frac{\square}{32}$$
$$\frac{7}{8} = \frac{7 \times 4}{8 \times 4}$$
$$\frac{7}{8} = \frac{28}{32}$$

$$7 = \frac{\square}{3}$$
$$\frac{7}{1} = \frac{7 \times 3}{1 \times 3}$$
$$7 = \frac{21}{3}$$

Name the whole number as a fraction whose denominator is 1. Then rename.

Rename.

	a	*b*	*c*
1.	$\frac{1}{2} = \frac{\square}{4}$	$\frac{1}{3} = \frac{\square}{9}$	$3 = \frac{\square}{12}$
2.	$6 = \frac{\square}{2}$	$\frac{4}{5} = \frac{\square}{10}$	$7 = \frac{\square}{5}$
3.	$\frac{1}{4} = \frac{\square}{8}$	$\frac{2}{3} = \frac{\square}{15}$	$4 = \frac{\square}{3}$
4.	$\frac{1}{3} = \frac{\square}{6}$	$\frac{1}{2} = \frac{\square}{8}$	$6 = \frac{\square}{6}$

NAME ______________________

Lesson 6 Addition (unlike denominators)

When adding fractions that have different denominators, rename the fractions so they have the same denominator.

$$\begin{array}{rcr} \frac{1}{3} & \times\frac{2}{2} & \frac{2}{6} \\ +\frac{1}{2} & \times\frac{3}{3} & +\frac{3}{6} \\ \hline & & \frac{5}{6} \end{array}$$

The denominators are 2 and 3. Since 2 × 3 = 6, rename each fraction with a denominator of 6.

Then add the fractions.

$$\begin{array}{rcr} \frac{1}{2} & \times\frac{3}{3} & \frac{3}{6} \\ +\frac{2}{3} & \times\frac{2}{2} & +\frac{4}{6} \\ \hline & & \frac{7}{6} = 1\frac{1}{6} \end{array}$$

Change $\frac{7}{6}$ to a mixed numeral in simplest form.

Write each answer in simplest form.

	a	*b*	*c*	*d*
1.	$\frac{2}{5} + \frac{1}{2}$	$\frac{1}{4} + \frac{2}{3}$	$\frac{2}{5} + \frac{1}{3}$	$\frac{1}{2} + \frac{1}{5}$
2.	$\frac{5}{6} + \frac{3}{5}$	$\frac{2}{3} + \frac{1}{5}$	$\frac{1}{3} + \frac{3}{10}$	$\frac{5}{8} + \frac{2}{3}$
3.	$\frac{3}{4} + \frac{1}{3}$	$\frac{2}{3} + \frac{4}{5}$	$\frac{2}{3} + \frac{3}{4}$	$\frac{7}{8} + \frac{1}{3}$

CHAPTER 12

Lesson 6 Addition

$$\begin{array}{r} \frac{2}{5} \\ +\frac{3}{10} \\ \hline \end{array} \quad \times\frac{2}{2} \longrightarrow \quad \begin{array}{r} \frac{4}{10} \\ +\frac{3}{10} \\ \hline \frac{7}{10} \end{array}$$

The denominators are 5 and 10. Since $2 \times 5 = 10$, rename only $\frac{2}{5}$ with a denominator of 10.

Then add the fractions.

$$\begin{array}{r} \frac{7}{10} \\ +\frac{2}{5} \\ \hline \end{array} \quad \times\frac{2}{2} \longrightarrow \quad \begin{array}{r} \frac{7}{10} \\ +\frac{4}{10} \\ \hline \frac{11}{10} \end{array} = 1\frac{1}{10}$$

Change $\frac{11}{10}$ to simplest form.

Write each answer in simplest form.

	a	*b*	*c*	*d*
1.	$\frac{3}{4} + \frac{1}{8}$	$\frac{2}{3} + \frac{5}{6}$	$\frac{1}{2} + \frac{3}{10}$	$\frac{5}{12} + \frac{2}{3}$
2.	$\frac{5}{16} + \frac{3}{8}$	$\frac{1}{6} + \frac{1}{2}$	$\frac{5}{8} + \frac{1}{4}$	$\frac{9}{10} + \frac{3}{5}$
3.	$\frac{3}{4} + \frac{9}{16}$	$\frac{5}{12} + \frac{1}{4}$	$\frac{5}{6} + \frac{1}{3}$	$\frac{1}{2} + \frac{7}{8}$

NAME ______________________

Lesson 7 Addition (unlike denominators)

$$\begin{array}{r} \frac{1}{6} \times \frac{4}{4} \quad \frac{4}{24} \\ +\frac{5}{8} \times \frac{3}{3} \quad +\frac{15}{24} \\ \hline \frac{19}{24} \end{array}$$

The denominators are 6 and 8. Since $4 \times 6 = 24$ and $3 \times 8 = 24$, rename each fraction with a denominator of 24.

Then add the fractions.

$$\begin{array}{r} \frac{5}{6} \times \frac{4}{4} \quad \frac{20}{24} \\ +\frac{3}{8} \times \frac{3}{3} \quad +\frac{9}{24} \\ \hline \frac{29}{24} = 1\frac{5}{24} \end{array}$$

Change $\frac{29}{24}$ to simplest form.

Write each answer in simplest form.

	a	*b*	*c*	*d*
1.	$\frac{1}{9} + \frac{1}{6}$	$\frac{1}{6} + \frac{1}{4}$	$\frac{5}{6} + \frac{1}{8}$	$\frac{1}{10} + \frac{1}{12}$
2.	$\frac{1}{6} + \frac{3}{8}$	$\frac{3}{4} + \frac{1}{6}$	$\frac{5}{6} + \frac{5}{8}$	$\frac{3}{10} + \frac{3}{8}$
3.	$\frac{3}{10} + \frac{5}{12}$	$\frac{5}{6} + \frac{4}{9}$	$\frac{3}{10} + \frac{1}{4}$	$\frac{5}{6} + \frac{3}{10}$
4.	$\frac{7}{10} + \frac{5}{6}$	$\frac{11}{12} + \frac{7}{8}$	$\frac{9}{10} + \frac{7}{8}$	$\frac{1}{4} + \frac{5}{6}$

CHAPTER 12

Lesson 7 Problem Solving

Solve. Write each answer in simplest form.

1. To make green paint, Andrea mixed $\frac{7}{8}$ quart of yellow paint and $\frac{1}{2}$ quart of blue paint. How much green paint did she make?

   She made ________ quarts of green paint.

2. Sean painted $\frac{1}{3}$ of a fence. Sandra painted $\frac{1}{4}$ of the fence. How much of the fence did they paint?

   They painted ________ of the fence.

3. Maureen bought $\frac{3}{4}$ pound of cheese. Chang bought $\frac{1}{2}$ pound of cheese. How much cheese did they buy?

   They bought ________ pounds of cheese.

4. A recipe calls for $\frac{2}{3}$ cup of milk and $\frac{3}{4}$ cup of water. How much milk and water are to be used?

   ________ cups of milk and water are to be used.

5. A board $\frac{1}{2}$ inch thick is glued to a board $\frac{3}{8}$ inch thick. What is the combined thickness?

   The combined thickness is ________ inch.

6. A book $\frac{3}{4}$ inch thick is placed on a book $\frac{13}{16}$ inch thick. What is the combined thickness of the books?

   The combined thickness is ________ inches.

7. Yesterday $\frac{3}{10}$ inch of rain fell. Today $\frac{3}{4}$ inch of rain fell. How much rain fell during the two days?

   ________ inches of rain fell during the two days.

1.	
2.	3.
4.	5.
6.	7.

NAME ____________________

Lesson 8 Addition (mixed numerals)

Rename the fractions so they have the same **denominator**.

$$\begin{array}{r} 3\frac{1}{4} \\ +2\frac{5}{6} \\ \hline \end{array} \rightarrow \begin{array}{r} 3\frac{3}{12} \\ +2\frac{10}{12} \\ \hline 5\frac{13}{12} \end{array} = 6\frac{1}{12}$$

$$\begin{array}{r} 4\frac{1}{2} \\ +3\frac{2}{3} \\ \hline \end{array} \rightarrow \begin{array}{r} 4\frac{3}{6} \\ +3\frac{4}{6} \\ \hline 7\frac{7}{6} \end{array} = 8\frac{1}{6}$$

Change to simplest form.

Write each answer in simplest form.

	a	*b*	*c*	*d*
1.	$3\frac{5}{6} + 4\frac{5}{8}$	$5\frac{2}{3} + 1\frac{5}{6}$	$6\frac{5}{6} + 3\frac{1}{4}$	$\frac{1}{2} + 2\frac{3}{4}$
2.	$1\frac{5}{6} + 4\frac{1}{3}$	$5\frac{1}{2} + 2\frac{3}{4}$	$3\frac{2}{3} + \frac{3}{4}$	$2\frac{3}{5} + 1\frac{1}{2}$
3.	$4\frac{3}{8} + 6\frac{1}{4}$	$5\frac{1}{3} + \frac{2}{5}$	$4\frac{2}{5} + 2\frac{3}{10}$	$2\frac{1}{8} + 5\frac{3}{4}$
4.	$3\frac{1}{2} + 3\frac{1}{2}$	$1\frac{3}{8} + 2\frac{1}{2}$	$9\frac{3}{4} + 6\frac{1}{2}$	$12\frac{2}{3} + 1\frac{5}{6}$

CHAPTER 12

Lesson 8 Problem Solving

Solve each problem.

1. Jennifer spent $1\frac{1}{2}$ hours working on Ms. Thomkin's car on Monday. She spent $2\frac{3}{4}$ more hours on Tuesday to finish the tune-up. How many hours in all did she work on Ms. Thomkin's car?

   She worked ________ hours in all.

   1.

2. Marissa worked $7\frac{1}{4}$ hours Monday. She worked $9\frac{3}{4}$ hours Tuesday. How many hours did she work in all on Monday and Tuesday?

   She worked ________ hours in all on Monday and Tuesday.

   2.

3. The auto repair shop is $1\frac{3}{10}$ miles from the bank. The bank is $3\frac{3}{5}$ miles from Gina's home. After she left her car at the shop, Gina walked to the bank. Then she walked home. How far did Gina walk in all?

   Gina walked ________ miles.

   3.

4. It took $2\frac{5}{6}$ hours to fix Mrs. Sax's car. It took $3\frac{1}{2}$ hours to fix Mr. Wong's car. How long did it take to fix both cars?

   It took ________ hours to fix both cars.

   4.

NAME ______________________

Lesson 9 Addition Review

Write each answer in simplest form.

	a	b	c	d
1.	$\frac{1}{12} + \frac{1}{6}$	$5\frac{5}{6} + 3\frac{5}{8}$	$4\frac{1}{3} + 2\frac{3}{4}$	$\frac{9}{16} + \frac{3}{4}$
2.	$1\frac{1}{4} + 6\frac{3}{5}$	$\frac{4}{7} + \frac{9}{10}$	$3\frac{3}{4} + \frac{9}{10}$	$\frac{7}{18} + \frac{7}{9}$
3.	$\frac{5}{7} + \frac{1}{2}$	$4\frac{2}{5} + 2\frac{8}{15}$	$\frac{5}{12} + 5\frac{3}{4}$	$\frac{9}{14} + \frac{3}{4}$
4.	$2\frac{1}{10} + 1\frac{1}{6}$	$\frac{1}{12} + \frac{5}{9}$	$\frac{5}{6} + \frac{1}{2}$	$8\frac{1}{3} + 3\frac{2}{9}$
5.	$\frac{2}{5} + \frac{3}{10}$	$\frac{7}{9} + 1\frac{1}{6}$	$5\frac{2}{5} + 3\frac{7}{10}$	$7\frac{3}{4} + 9\frac{5}{6}$

CHAPTER 12

Lesson 9 Problem Solving

Solve. Write each answer in simplest form.

1. Emilio weighs $71\frac{1}{4}$ pounds. His sister weighs $10\frac{3}{4}$ pounds more than that. How much does his sister weigh?

   His sister weighs ________ pounds.

2. Arlene spent $2\frac{1}{2}$ hours planting part of a garden. It took her $1\frac{3}{4}$ hours to finish planting the garden. How long did it take to plant the garden?

   It took ________ hours.

3. A basket weighs $1\frac{1}{8}$ pounds when empty. Jake put $10\frac{1}{2}$ pounds of apples in the basket. How much do the basket and apples weigh?

   The basket and apples weigh ________ pounds.

4. Normal body temperature is $98\frac{6}{10}$°F. The doctor said June's temperature is $2\frac{1}{2}$ degrees above normal. What is her temperature?

   Her temperature is ________ °F.

5. Ned jumped a distance of $4\frac{1}{3}$ feet. Phil jumped $1\frac{1}{4}$ feet farther than Ned. How far did Phil jump?

   Phil jumped ________ feet.

6. A board $1\frac{3}{8}$ inches thick is glued to a board $1\frac{3}{4}$ inches thick. What is the combined thickness of the boards?

   The combined thickness is ________ inches.

1.	2.
3.	4.
5.	6.

NAME ______________________

Lesson 10 Addition Review

Write each answer in simplest form.

	a	*b*	*c*	*d*
1.	$\frac{1}{9} + \frac{4}{9}$	$\frac{2}{7} + \frac{3}{7}$	$\frac{8}{9} + \frac{5}{9}$	$\frac{11}{16} + \frac{7}{16}$
2.	$\frac{2}{3} + \frac{1}{5}$	$\frac{2}{5} + \frac{3}{4}$	$\frac{1}{2} + \frac{3}{4}$	$\frac{5}{6} + \frac{1}{12}$
3.	$\frac{7}{8} + \frac{5}{6}$	$\frac{5}{12} + \frac{1}{3}$	$\frac{1}{5} + \frac{7}{10}$	$\frac{7}{8} + \frac{5}{12}$
4.	$\frac{2}{5} + \frac{1}{5}$	$2\frac{1}{9} + \frac{1}{3}$	$7\frac{5}{8} + \frac{2}{3}$	$4\frac{7}{12} + 1\frac{1}{2}$
5.	$\frac{1}{5} + \frac{1}{3}$	$\frac{3}{4} + \frac{1}{5}$	$1\frac{2}{3} + 1\frac{5}{6}$	$3\frac{11}{12} + 2\frac{5}{6}$

CHAPTER 12

Lesson 10 Problem Solving

Solve. Write each answer in simplest form.

1. Jared lives $\frac{7}{8}$ mile from the stadium and $\frac{3}{8}$ mile from the school. He walked home from school and then to the stadium. How far did he walk?

Jared walked ________ miles.

2. Courtney read $\frac{5}{6}$ hour before dinner. After dinner she read $\frac{2}{5}$ hour. How long did she read?

Courtney read ________ hours in all.

3. The Clements family drank $\frac{3}{4}$ gallon of milk for dinner. There was $\frac{1}{2}$ gallon left. How much milk was there before dinner?

There were ________ gallons of milk.

4. Gary rides the bus $1\frac{3}{10}$ miles every day. Glen rides $\frac{3}{10}$ mile farther than Gary. How far does Glen ride?

Glen rides ________ miles every day.

5. Rocio is $4\frac{3}{4}$ feet tall. Her father is $1\frac{1}{2}$ feet taller than that. How tall is Rocio's father?

He is ________ feet tall.

6. To make pale blue paint, Lynn mixed $2\frac{1}{4}$ gallons of blue paint and $3\frac{3}{4}$ gallons of white paint. How much pale blue paint did she make?

She made ________ gallons of pale blue paint.

7. Last year Becky was $49\frac{1}{2}$ inches tall. Since then she has grown $1\frac{7}{8}$ inches. How tall is she now?

She is now ________ inches tall.

1.	
2.	3.
4.	5.
6.	7.

NAME ______________________

CHAPTER 12 PRACTICE TEST
Addition of Fractions

Write each answer in simplest form.

	a	*b*	*c*	*d*
1.	$\frac{3}{10} + \frac{1}{10}$	$\frac{5}{6} + \frac{1}{6}$	$\frac{7}{8} + \frac{5}{8}$	$\frac{4}{7} + \frac{1}{7}$
2.	$\frac{5}{8} + \frac{1}{4}$	$\frac{3}{10} + \frac{3}{4}$	$\frac{1}{2} + \frac{4}{5}$	$\frac{5}{6} + \frac{3}{4}$
3.	$5\frac{3}{10} + 1\frac{1}{3}$	$4\frac{2}{9} + 2\frac{2}{3}$	$\frac{5}{6} + 3\frac{1}{12}$	$6\frac{5}{12} + \frac{1}{3}$
4.	$1\frac{3}{4} + 4\frac{7}{10}$	$5\frac{1}{3} + \frac{4}{5}$	$2\frac{3}{4} + 6\frac{15}{16}$	$7\frac{7}{10} + 8\frac{4}{5}$
5.	$7\frac{1}{5} + \frac{1}{4}$	$9\frac{9}{10} + \frac{7}{12}$	$42\frac{5}{6} + 5\frac{2}{3}$	$54\frac{1}{2} + 21\frac{4}{5}$

CHAPTER 12

NAME ____________________

CHAPTER 13 PRETEST
Subtraction of Fractions

Write each answer in simplest form.

	a	*b*	*c*	*d*
1.	$\frac{7}{8} - \frac{3}{8}$	$\frac{8}{9} - \frac{2}{9}$	$\frac{5}{6} - \frac{1}{6}$	$\frac{11}{12} - \frac{3}{12}$
2.	$5\frac{4}{5} - 2\frac{1}{5}$	$4\frac{5}{9} - 3\frac{2}{9}$	$6\frac{4}{7} - 1\frac{6}{7}$	$3\frac{3}{8} - \frac{7}{8}$
3.	$\frac{5}{6} - \frac{2}{3}$	$\frac{2}{3} - \frac{1}{2}$	$\frac{8}{9} - \frac{1}{3}$	$\frac{7}{8} - \frac{3}{4}$
4.	$\frac{7}{10} - \frac{1}{5}$	$\frac{7}{8} - \frac{3}{10}$	$\frac{9}{10} - \frac{2}{5}$	$\frac{5}{6} - \frac{7}{12}$
5.	$4\frac{5}{6} - 2\frac{1}{3}$	$3\frac{7}{8} - 1\frac{2}{3}$	$2\frac{1}{10} - 1\frac{4}{5}$	$2\frac{1}{5} - \frac{2}{3}$

CHAPTER 13

NAME ____________________

Lesson 1 Subtraction (like denominators)

Study how to subtract when fractions have the same denominator.

Subtract the numerators.

$\frac{7}{8} - \frac{5}{8} = \frac{7-5}{8} = \frac{2}{8} = \frac{1}{4}$

Use the same denominator. Change to simplest form.

Subtract the numerators. Use the same denominator.

$$\begin{array}{r} \frac{7}{8} \\ -\frac{5}{8} \\ \hline \frac{2}{8} = \frac{1}{4} \end{array}$$

Change to simplest form.

Write each answer in simplest form.

	a	*b*	*c*	*d*	*e*
1.	$\frac{5}{9} - \frac{4}{9}$	$\frac{3}{5} - \frac{1}{5}$	$\frac{8}{9} - \frac{4}{9}$	$\frac{3}{4} - \frac{1}{4}$	$\frac{5}{6} - \frac{1}{6}$
2.	$\frac{6}{7} - \frac{4}{7}$	$\frac{5}{8} - \frac{3}{8}$	$\frac{9}{10} - \frac{3}{10}$	$\frac{2}{5} - \frac{1}{5}$	$\frac{5}{9} - \frac{1}{9}$
3.	$\frac{5}{7} - \frac{2}{7}$	$\frac{8}{9} - \frac{1}{9}$	$\frac{7}{8} - \frac{3}{8}$	$\frac{7}{12} - \frac{5}{12}$	$\frac{9}{10} - \frac{7}{10}$
4.	$\frac{4}{5} - \frac{2}{5}$	$\frac{2}{3} - \frac{1}{3}$	$\frac{7}{10} - \frac{3}{10}$	$\frac{7}{9} - \frac{4}{9}$	$\frac{7}{8} - \frac{1}{8}$

CHAPTER 13

NAME ______________________

Lesson 2 Subtraction (from whole numbers)

Rename the whole number as a mixed numeral so the denominator is the same as that of the fraction.

$$\begin{array}{r} 2 \\ -\frac{3}{4} \\ \hline \end{array} \rightarrow \begin{array}{r} 1\frac{4}{4} \\ -\frac{3}{4} \\ \hline 1\frac{1}{4} \end{array}$$

$2 = 1 + 1$
$= 1 + \frac{4}{4}$
$= 1\frac{4}{4}$

$$\begin{array}{r} 5 \\ -\frac{7}{8} \\ \hline \end{array} \rightarrow \begin{array}{r} 4\frac{8}{8} \\ -\frac{7}{8} \\ \hline 4\frac{1}{8} \end{array}$$

$5 = 4 + 1$
$= 4 + \frac{8}{8}$
$= 4\frac{8}{8}$

Write each answer in simplest form.

	a	*b*	*c*	*d*
1.	$2 - \frac{1}{4}$	$3 - \frac{2}{3}$	$6 - \frac{1}{5}$	$5 - \frac{1}{3}$
2.	$4 - \frac{3}{4}$	$5 - \frac{2}{5}$	$4 - \frac{2}{5}$	$6 - \frac{5}{6}$
3.	$1 - \frac{1}{2}$	$2 - \frac{7}{8}$	$1 - \frac{1}{8}$	$2 - \frac{3}{10}$

NAME ______________________

Lesson 3 Subtraction (mixed numerals)

$\frac{1}{4}$ is less than $\frac{3}{4}$. So rename $7\frac{1}{4}$ as shown so you can subtract the fractions.

$$\begin{array}{r} 7\frac{1}{4} \\ -1\frac{3}{4} \\ \hline \end{array} \rightarrow \begin{array}{r} 6\frac{5}{4} \\ -1\frac{3}{4} \\ \hline 5\frac{2}{4} \end{array} = 5\frac{1}{2}$$

Change to simplest form.

$$\begin{aligned} 7\tfrac{1}{4} &= 6 + 1 + \tfrac{1}{4} \\ &= 6 + \tfrac{4}{4} + \tfrac{1}{4} \\ &= 6\tfrac{5}{4} \end{aligned}$$

$\frac{1}{3}$ is less than $\frac{2}{3}$. So rename $3\frac{1}{3}$ as shown so you can subtract the fractions.

$$\begin{array}{r} 3\frac{1}{3} \\ -2\frac{2}{3} \\ \hline \end{array} \rightarrow \begin{array}{r} 2\frac{4}{3} \\ -2\frac{2}{3} \\ \hline \frac{2}{3} \end{array}$$

$$\begin{aligned} 3\tfrac{1}{3} &= 2 + 1 + \tfrac{1}{3} \\ &= 2 + \tfrac{3}{3} + \tfrac{1}{3} \\ &= 2\tfrac{4}{3} \end{aligned}$$

Write each answer in simplest form.

	a	*b*	*c*	*d*
1.	$5\frac{8}{9} - 2\frac{6}{9}$	$4\frac{6}{7} - 2\frac{1}{7}$	$8\frac{9}{10} - 3\frac{4}{10}$	$6\frac{3}{8} - 2\frac{1}{8}$
2.	$5\frac{1}{3} - 1\frac{2}{3}$	$7\frac{2}{5} - 1\frac{4}{5}$	$8\frac{3}{8} - 2\frac{5}{8}$	$6\frac{1}{9} - 2\frac{6}{9}$
3.	$5\frac{3}{12} - 2\frac{11}{12}$	$4\frac{5}{6} - 2\frac{2}{6}$	$3\frac{2}{5} - 1\frac{4}{5}$	$7\frac{2}{3} - 6\frac{2}{3}$

CHAPTER 13

Lesson 3 Problem Solving

Solve. Write each answer in simplest form.

1. A board is 8 feet long. Hank said that this board is $2\frac{1}{2}$ feet too long for the job. How long is the board that Hank needs?

   He needs a board ________ feet long.

2. Sue says it will take $6\frac{1}{6}$ hours to travel to her grandparents' home. She has been traveling $3\frac{5}{6}$ hours. How much longer will it be before she gets there?

   It will be ________ hours longer.

3. The stakes in Don's croquet set are 2 feet long. He drove one stake $\frac{3}{4}$ foot into the ground. How much of the stake is above the ground?

   ________ feet are above the ground.

4. An envelope is 7 inches wide. A sheet of paper is $6\frac{1}{2}$ inches wide. How much wider than the paper is the envelope?

   It is ________ inch wider.

5. This year Reola spends $5\frac{1}{4}$ hours in school each day. Last year she spent $4\frac{3}{4}$ hours in school each day. How many more hours does she spend in school each day this year than last year?

   She spends ________ hour more in school each day this year than last year.

6. A wire is $4\frac{7}{12}$ feet long. Suppose $\frac{11}{12}$ foot of wire is used. How much wire would be left?

   ________ feet of wire would be left.

1.	2.
3.	4.
5.	6.

NAME ______________________

Lesson 4 Subtraction (unlike denominators)

When subtracting fractions that have different denominators, rename the fractions so they have the same denominator.

$$\begin{array}{r} \frac{2}{3} \times \frac{4}{4} = \frac{8}{12} \\ -\frac{1}{4} \times \frac{3}{3} = -\frac{3}{12} \\ \hline \frac{5}{12} \end{array}$$

Since 3 × 4 = 12, rename each fraction with a denominator of 12. Then subtract.

$$\begin{array}{r} \frac{5}{6} \longrightarrow \frac{5}{6} \\ -\frac{1}{2} \times \frac{3}{3} = -\frac{3}{6} \\ \hline \frac{2}{6} = \frac{1}{3} \end{array}$$

Since 2 × 3 = 6, rename only $\frac{1}{2}$ with a denominator of 6. Then subtract.

Write each answer in simplest form.

	a	*b*	*c*	*d*
1.	$\frac{3}{5} - \frac{1}{3}$	$\frac{5}{6} - \frac{2}{5}$	$\frac{7}{8} - \frac{1}{2}$	$\frac{2}{3} - \frac{4}{9}$
2.	$\frac{5}{6} - \frac{1}{3}$	$\frac{2}{3} - \frac{1}{6}$	$\frac{7}{12} - \frac{1}{4}$	$\frac{4}{5} - \frac{3}{10}$
3.	$\frac{9}{10} - \frac{1}{2}$	$\frac{5}{6} - \frac{3}{7}$	$\frac{3}{4} - \frac{1}{5}$	$\frac{11}{12} - \frac{1}{6}$

CHAPTER 13

Lesson 4 Problem Solving

Solve. Write each answer in simplest form.

1. Phillip jogged $\frac{5}{6}$ mile. He walked $\frac{1}{2}$ mile. How much farther did he jog than he walked?

   He jogged ________ mile farther than he walked.

2. Kyle and Eric have painted $\frac{2}{3}$ of a room. Kyle painted $\frac{1}{2}$ of the room. How much of the room did Eric paint?

   Eric painted ________ of the room.

3. Rona and Joan have $\frac{5}{6}$ of a room painted. Joan painted $\frac{1}{5}$ of the room. How much of the room did Rona paint?

   Rona painted ________ of the room.

4. Ardith had $\frac{3}{4}$ dozen eggs. She used $\frac{7}{12}$ dozen for breakfast. How many dozen does she have left?

   She has ________ dozen eggs left.

5. A rock weighs $\frac{9}{16}$ pound. Suppose $\frac{1}{4}$ pound is chipped away. How much would the remaining rock weigh?

   The remaining part would weigh ________ pound.

6. It takes Monica $\frac{5}{6}$ hour to get to work. In doing so, she rides the train $\frac{2}{3}$ hour. She walks the remaining time. How much time does she spend walking to work?

   She spends ________ hour walking to work.

7. Mr. Anthony and Mr. Androtti completed $\frac{3}{4}$ of a job. Mr. Androtti completed $\frac{2}{9}$ of the job. What part of the job did Mr. Anthony complete?

   Mr. Anthony completed ________ of the job.

1.
2.
3.
4.
5.
6.
7.

NAME ____________________

Lesson 5 Subtraction (unlike denominators)

$$\begin{array}{r} \frac{3}{4} \\ -\frac{3}{5} \\ \hline \end{array} \rightarrow \begin{array}{r} \frac{15}{20} \\ -\frac{12}{20} \\ \hline \frac{3}{20} \end{array} \qquad \begin{array}{r} \frac{9}{10} \\ -\frac{11}{15} \\ \hline \end{array} \rightarrow \begin{array}{r} \frac{27}{30} \\ -\frac{22}{30} \\ \hline \frac{5}{30} = \frac{1}{6} \end{array}$$

Write each answer in simplest form.

	a	*b*	*c*	*d*
1.	$\frac{5}{6} - \frac{3}{8}$	$\frac{3}{4} - \frac{1}{6}$	$\frac{7}{8} - \frac{3}{10}$	$\frac{5}{6} - \frac{2}{9}$
2.	$\frac{9}{10} - \frac{3}{5}$	$\frac{7}{8} - \frac{1}{6}$	$\frac{2}{3} - \frac{1}{5}$	$\frac{8}{9} - \frac{5}{6}$
3.	$\frac{3}{4} - \frac{5}{12}$	$\frac{7}{12} - \frac{1}{4}$	$\frac{7}{8} - \frac{1}{3}$	$\frac{3}{10} - \frac{1}{4}$
4.	$\frac{2}{3} - \frac{4}{9}$	$\frac{11}{12} - \frac{3}{8}$	$\frac{1}{4} - \frac{1}{12}$	$\frac{2}{3} - \frac{7}{12}$

Lesson 5 Problem Solving

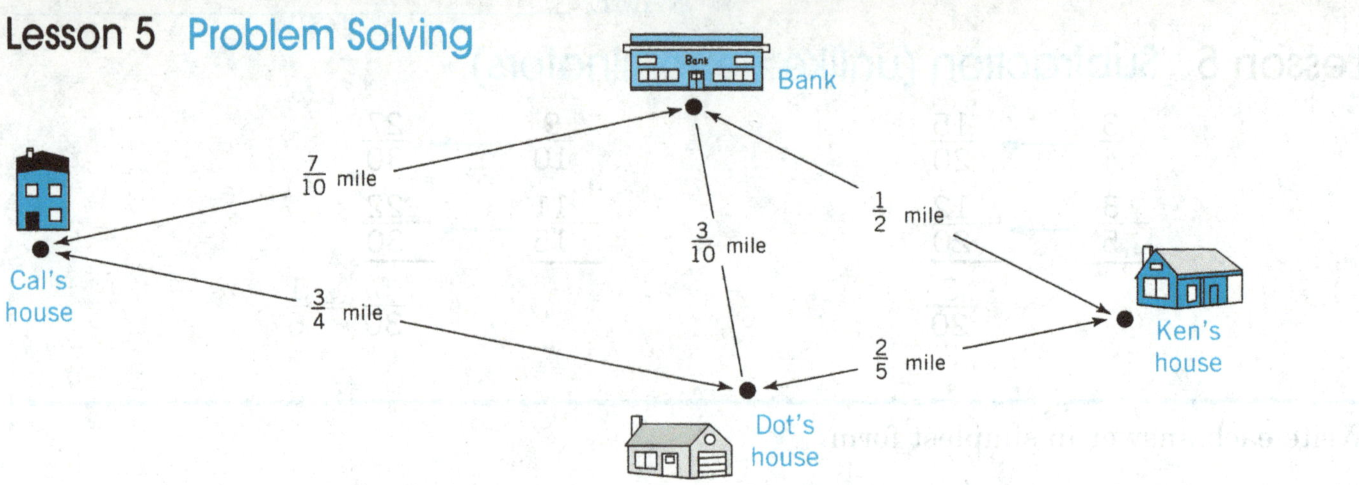

Solve. Write each answer in simplest form.

1. Who lives farther from the bank, Cal or Dot? How much farther?

   ________ lives ________ mile farther.

2. Who lives farther from the bank, Ken or Cal? How much farther?

   ________ lives ________ mile farther.

3. How much farther is it from Dot's house to Cal's house than from Dot's house to the bank?

   It is ________ mile farther.

4. How much farther is it from Dot's house to Ken's house than from Dot's house to the bank?

   It is ________ mile farther.

5. Cal walked from his house to Dot's house. Ken walked from his house to Dot's house. Who walked farther? How much farther?

   ________ walked ________ mile farther.

1.	
2.	3.
4.	5.

NAME ______________________

Lesson 6 Subtraction (mixed numerals)

Rename so the fractions have the same denominator.

Rename $7\frac{3}{12}$ so you can subtract.

$$\begin{array}{r} 7\frac{1}{4} \\ -3\frac{2}{3} \\ \hline \end{array} \rightarrow \begin{array}{r} 7\frac{3}{12} \\ -3\frac{8}{12} \\ \hline \end{array} \rightarrow \begin{array}{r} 6\frac{15}{12} \\ -3\frac{8}{12} \\ \hline 3\frac{7}{12} \end{array}$$

$$\begin{aligned} 7\tfrac{3}{12} &= 7 + \tfrac{3}{12} \\ &= 6 + 1 + \tfrac{3}{12} \\ &= 6 + \tfrac{12}{12} + \tfrac{3}{12} \\ &= 6 + \tfrac{15}{12} \\ &= 6\tfrac{15}{12} \end{aligned}$$

Write each answer in simplest form.

	a	*b*	*c*	*d*
1.	$5\frac{1}{3} - 3\frac{3}{4}$	$7\frac{3}{5} - 4\frac{7}{10}$	$6\frac{1}{6} - 1\frac{3}{8}$	$5\frac{4}{9} - 2\frac{1}{3}$
2.	$4\frac{3}{8} - 2\frac{1}{3}$	$3\frac{5}{6} - 2\frac{1}{12}$	$6\frac{4}{7} - 5\frac{1}{2}$	$6\frac{3}{5} - 2\frac{3}{10}$
3.	$5\frac{7}{8} - 1\frac{3}{5}$	$3\frac{1}{9} - \frac{1}{3}$	$2\frac{2}{3} - 1\frac{1}{2}$	$1\frac{3}{8} - \frac{9}{10}$
4.	$4\frac{2}{9} - \frac{2}{3}$	$6\frac{4}{5} - 5\frac{3}{7}$	$3\frac{7}{12} - 1\frac{9}{10}$	$2\frac{1}{8} - \frac{5}{12}$

CHAPTER 13

Lesson 6 Problem Solving

Solve. Write each answer in simplest form.

1. One fish weighed $1\frac{1}{2}$ pounds. Another weighed $\frac{3}{4}$ pound. How much more did the heavier fish weigh?

   It weighed ________ pound more.

2. Mrs. Tanner bought $2\frac{1}{2}$ gallons of paint. She used $1\frac{2}{3}$ gallons of paint on the garage. How much paint did she have left?

   She had ________ gallon left.

3. Lorena has two boxes that weigh a total of $4\frac{1}{2}$ pounds. One box weighs $1\frac{7}{10}$ pounds. How much does the other box weigh?

   It weighs ________ pounds.

4. Allen practiced the guitar $1\frac{1}{4}$ hours today. He practiced $\frac{2}{3}$ hour before lunch. How long did he practice after lunch?

   He practiced________ hour after lunch.

5. Karen ran a race in $9\frac{3}{10}$ seconds. Curt ran the race in $7\frac{4}{5}$ seconds. How much longer did it take Karen to run the race?

   It took ________ seconds longer.

6. Fido weighs $2\frac{5}{16}$ pounds. Spot weighs $4\frac{7}{8}$ pounds. How much more does Spot weigh?

   Spot weighs ________ pounds more.

1.
2.
3.
4.
5.
6.

NAME ____________________

Lesson 7 Subtraction Review

Write each answer in simplest form.

	a	*b*	*c*	*d*
1.	$\frac{7}{9} - \frac{4}{9}$	$\frac{7}{8} - \frac{1}{2}$	$\frac{7}{8} - \frac{3}{16}$	$\frac{11}{12} - \frac{1}{6}$
2.	$\frac{4}{5} - \frac{2}{3}$	$\frac{7}{10} - \frac{6}{10}$	$\frac{9}{10} - \frac{2}{5}$	$\frac{11}{12} - \frac{3}{4}$
3.	$\frac{5}{12} - \frac{3}{12}$	$\frac{3}{8} - \frac{1}{5}$	$\frac{5}{8} - \frac{3}{8}$	$\frac{2}{3} - \frac{1}{6}$
4.	$4\frac{7}{10} - 1\frac{2}{5}$	$3\frac{5}{12} - 1\frac{1}{12}$	$8\frac{3}{10} - 5\frac{9}{10}$	$5\frac{3}{8} - 3\frac{5}{8}$
5.	$1\frac{1}{4} - \frac{3}{10}$	$4\frac{6}{7} - 2\frac{3}{7}$	$1\frac{1}{3} - \frac{5}{6}$	$2\frac{4}{5} - \frac{9}{10}$

CHAPTER 13

Lesson 7 Problem Solving

Solve. Write each answer in simplest form.

1. A pail filled with water weighs $9\frac{1}{4}$ pounds. The empty pail weighs $\frac{3}{4}$ pound. How much does the water weigh?

   The water weighs ________ pounds.

2. A board is $4\frac{5}{8}$ inches long. We need a piece $2\frac{7}{8}$ inches long. How much of the board needs to be cut?

   ________ inches need to be cut.

3. John and Mara are reading the same book. John has read $\frac{4}{5}$ of the book and Mara has read $\frac{2}{3}$ of the book. How much more of the book has John read than Mara?

   John has read ________ more of the book.

4. A recipe calls for $3\frac{1}{2}$ cups of flour and $1\frac{3}{4}$ cups of sugar. How many more cups of flour than sugar are called for by the recipe?

   ________ cups more of flour are called for.

5. Meagan worked $7\frac{1}{2}$ hours. Joshua worked $5\frac{3}{4}$ hours. How much longer than Joshua did Meagan work?

   She worked ________ hours longer.

6. It took Amber $2\frac{2}{3}$ hours to read 2 books. She read one book in $\frac{5}{6}$ hour. How long did it take her to read the other one?

   It took ________ hours to read the other book.

7. Mr. Wakefield used $8\frac{1}{4}$ gallons of water to fill two tanks. He put $3\frac{7}{8}$ gallons in one tank. How much water did he put in the other tank?

   He put ________ gallons in the other tank.

1.
2.
3.
4.
5.
6.
7.

NAME ____________________

Lesson 8 Subtraction Review

Write each answer in simplest form.

	a	*b*	*c*	*d*
1.	$\frac{7}{9} - \frac{2}{9}$	$\frac{5}{7} - \frac{1}{7}$	$\frac{5}{8} - \frac{3}{8}$	$\frac{7}{10} - \frac{1}{10}$
2.	$3\frac{5}{6} - 2\frac{1}{6}$	$4\frac{5}{9} - 3\frac{2}{9}$	$5\frac{1}{4} - 1\frac{3}{4}$	$1\frac{4}{15} - \frac{7}{15}$
3.	$\frac{3}{4} - \frac{2}{3}$	$\frac{4}{5} - \frac{2}{3}$	$\frac{3}{4} - \frac{1}{2}$	$\frac{5}{9} - \frac{1}{3}$
4.	$\frac{7}{8} - \frac{3}{4}$	$\frac{5}{6} - \frac{1}{2}$	$\frac{3}{4} - \frac{1}{6}$	$\frac{7}{10} - \frac{1}{12}$
5.	$3\frac{7}{8} - 2\frac{1}{6}$	$4\frac{7}{10} - 1\frac{4}{5}$	$5\frac{5}{12} - 3\frac{7}{10}$	$6\frac{2}{9} - \frac{11}{12}$

CHAPTER 13

Lesson 8 Problem Solving

Animal	*Weight*
dog	$4\frac{1}{2}$ lbs
cat	$2\frac{2}{3}$ lbs
rabbit	$1\frac{3}{4}$ lbs

Solve. Write each answer in simplest form.

1. How much more does the dog weigh than the cat?

   The dog weighs ________ pounds more than the cat.

2. How much more does the dog weigh than the rabbit?

   The dog weighs ________ pounds more than the rabbit.

3. How much more does the cat weigh than the rabbit?

   The cat weighs ________ pound more than the rabbit.

4. How much do the dog and the cat weigh together?

   Together, the dog and the cat weigh ________ pounds.

1.	2.
3.	4.

NAME ______________________

CHAPTER 13 PRACTICE TEST
Subtraction of Fractions

Write each answer in simplest form.

	a	*b*	*c*	*d*
1.	$\frac{9}{10} - \frac{7}{10}$	$\frac{4}{5} - \frac{2}{3}$	$\frac{3}{4} - \frac{5}{8}$	$\frac{8}{9} - \frac{2}{9}$
2.	$\frac{5}{6} - \frac{1}{2}$	$\frac{1}{2} - \frac{3}{8}$	$\frac{11}{12} - \frac{3}{12}$	$\frac{1}{2} - \frac{5}{12}$
3.	$\frac{3}{4} - \frac{3}{8}$	$\frac{5}{6} - \frac{1}{9}$	$\frac{7}{8} - \frac{1}{4}$	$\frac{2}{3} - \frac{1}{2}$
4.	$5\frac{7}{8} - 2\frac{3}{8}$	$4\frac{2}{5} - 2\frac{3}{10}$	$6\frac{1}{2} - 1\frac{1}{3}$	$3\frac{1}{3} - 1\frac{5}{6}$
5.	$3\frac{11}{12} - 1\frac{5}{6}$	$5\frac{5}{8} - 2\frac{3}{4}$	$2\frac{1}{9} - \frac{7}{9}$	$1\frac{2}{5} - \frac{1}{2}$

CHAPTER 13

CHAPTER 14 PRETEST
Geometry

Circle the phrase that correctly describes each figure.

1. line *MN*   line segment *MN*   line *M*

2. line *R*   line segment *RP*   line *RP*

3. line segment *JK* ∥ line segment *LM*
   line *JK* ⊥ line *LM*
   line segment *JK* ⊥ line segment *LM*

4. line *DE* ∥ line *FG*
   line *DE* ⊥ line *FG*
   line segment *DE* ∥ line segment *FG*

Name each angle. Give its measure and identify it as *acute*, *obtuse*, or *right*.

5. *a*

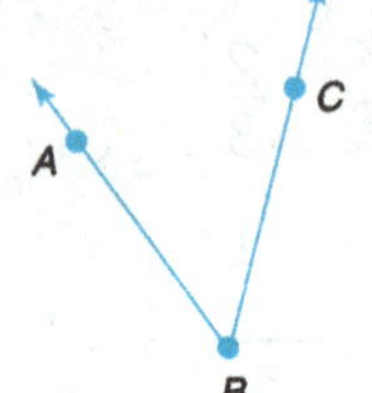

b

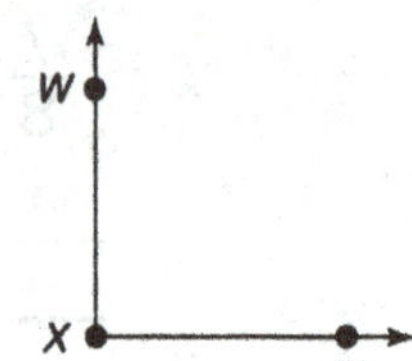

c

______________   ______________   ______________

______________   ______________   ______________

Write the letter for the name of each figure in the blank.

a   *b*

6. ______ 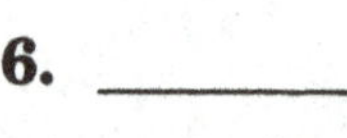  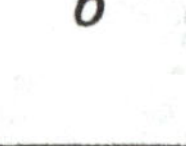 ______ 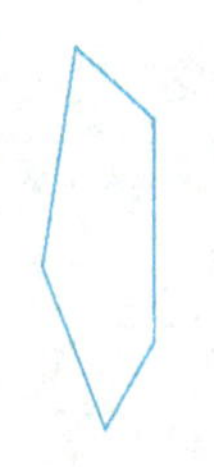

7. ______   ______

a. octagon
b. triangle
c. hexagon
d. pentagon
e. square
f. quadrilateral
g. circle

Lesson 1 Lines and Line Segments

A **line** has no endpoints.

To name a line, name any two points on the line.

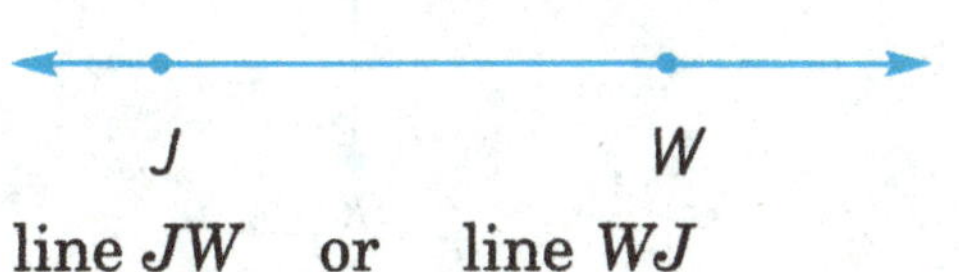

line *JW* or line *WJ*

A **line segment** has two **endpoints.**

A line segment is part of a line. The line segment consists of the endpoints and all points on the line between the endpoints. To name a line segment, name the endpoints.

line segment *GS* or line segment *SG*

Circle the correct name for each figure.

Figure			
1. line with points *B*, *A*	line *AB*	line segment *BA*	line *CA*
2. segment with endpoints *G*, *F*	line segment *FG*	line *GF*	line *FG*
3. segment with endpoints *C*, *E*	line *CD*	line segment *CE*	line *CE*
4. line with points *M*, *N*	line segment *MN*	line *MM*	line *MN*
5. segment with endpoints *R*, *S*	line *RS*	line segment *RS*	line *SR*
6. segment with endpoints *K*, *I*	line segment *KI*	line *KI*	line *IK*
7. line with points *Z*, *X*	line *LZ*	line segment *ZX*	line *ZX*
8. line with points *E*, *P*	line segment *PE*	line *EP*	line *EE*
9. segment with endpoints *T*, *V*	line *V*	line segment *VT*	line *VT*

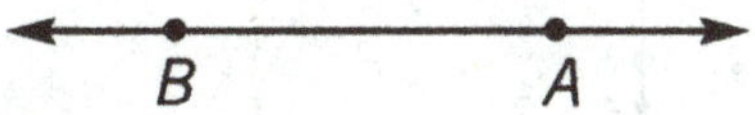

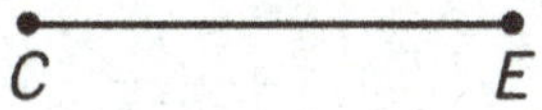

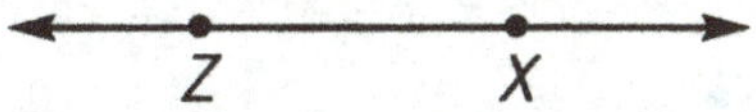

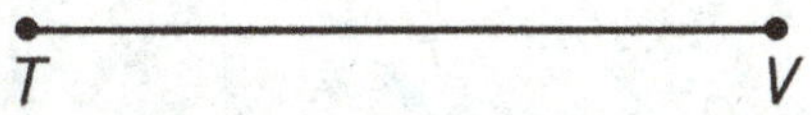

Draw and label the following.

10. line segment *HQ*

CHAPTER 14

NAME ____________________

Lesson 2 Parallel and Perpendicular

Parallel lines or line segments never intersect and are always the same distance apart. Parallel is indicated by the symbol ∥.

Use symbols to identify each figure.

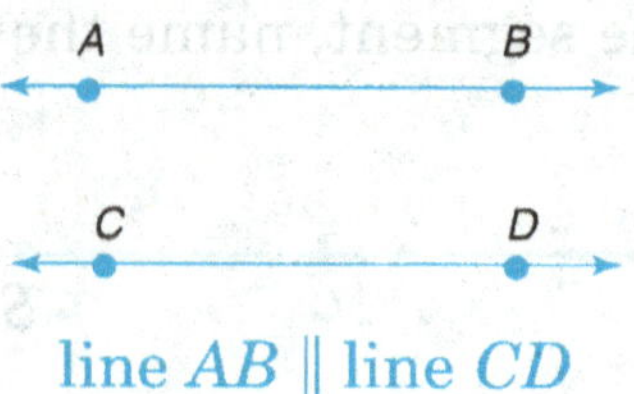

line *AB* ∥ line *CD*

Perpendicular lines or line segments intersect each other and form 90° angles. Perpendicular is indicated by the symbol ⊥.

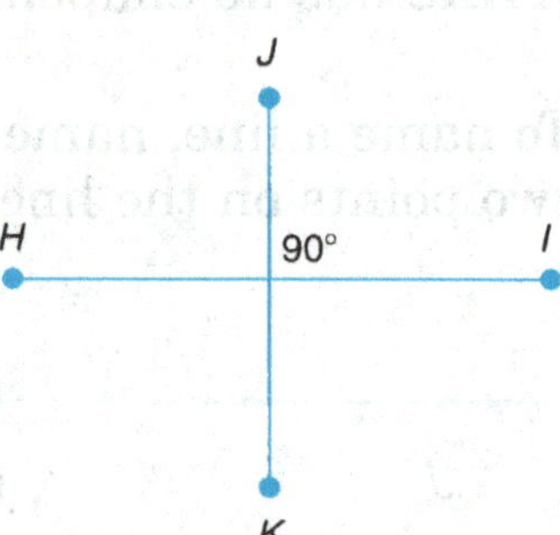

line segment *HI* ⊥ line segment *JK*

Use symbols to identify each figure.

a *b*

1. 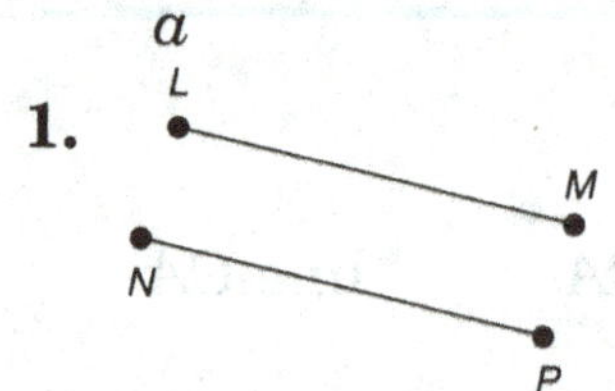

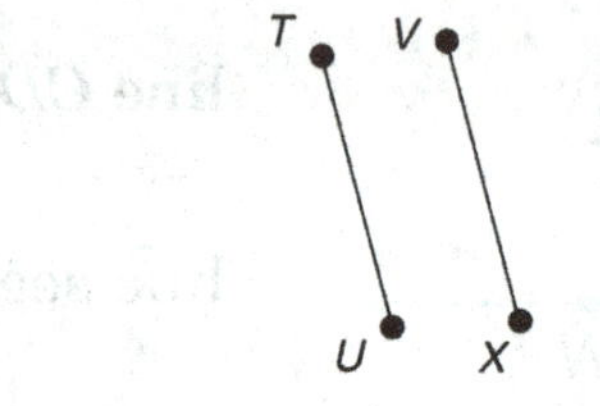

______________________ ______________________

2. 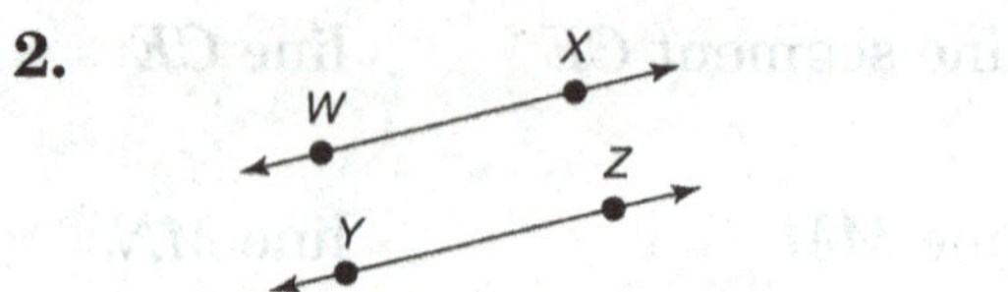

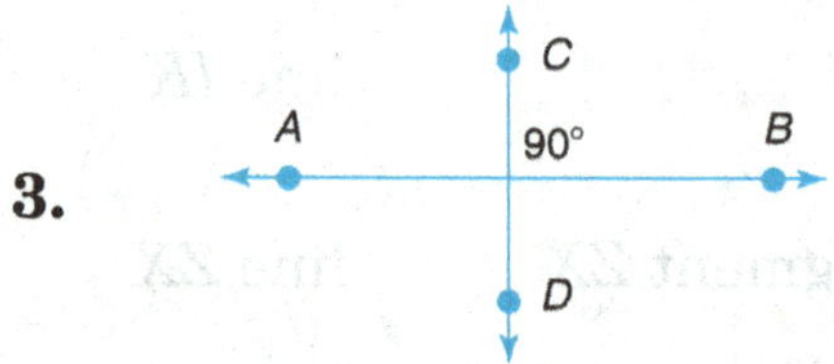

______________________ ______________________

3. 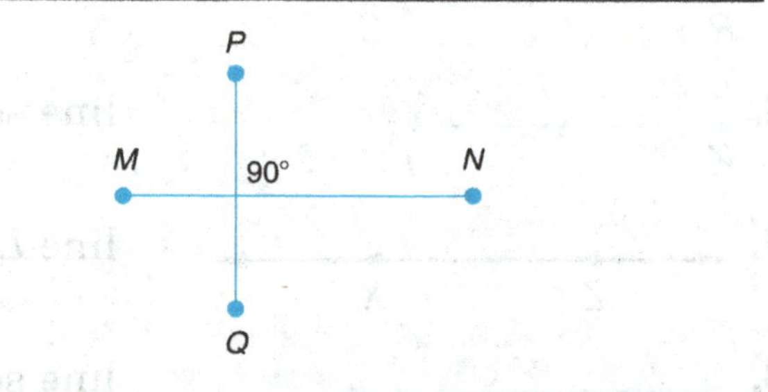

______________________ ______________________

4. 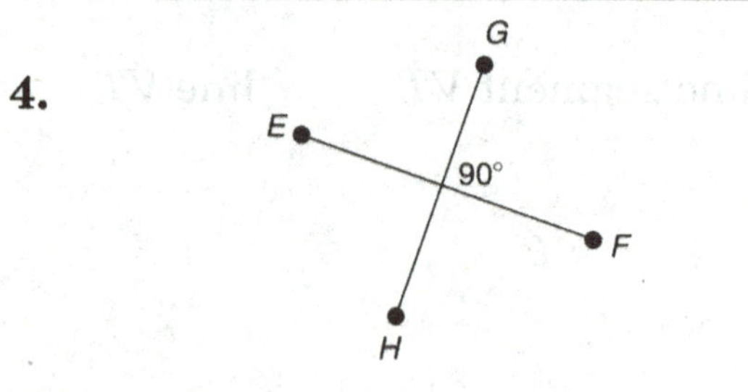

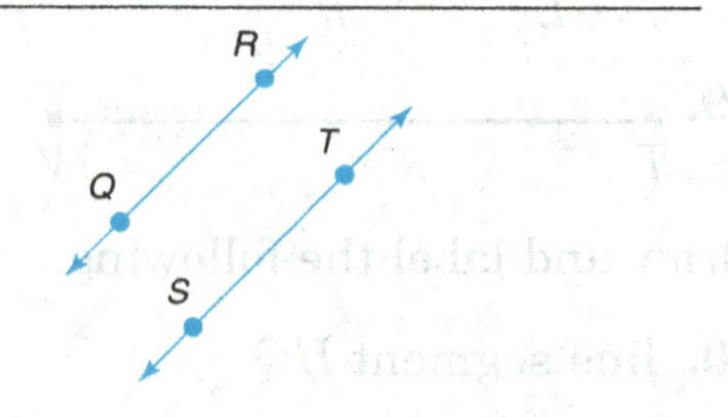

______________________ ______________________

Lesson 3 Angles

An **angle** has two sides and a **vertex.**

Angle *GHB* (denoted ∠*GHB*) has a vertex of *H*. When naming an angle, use the vertex as the middle letter.

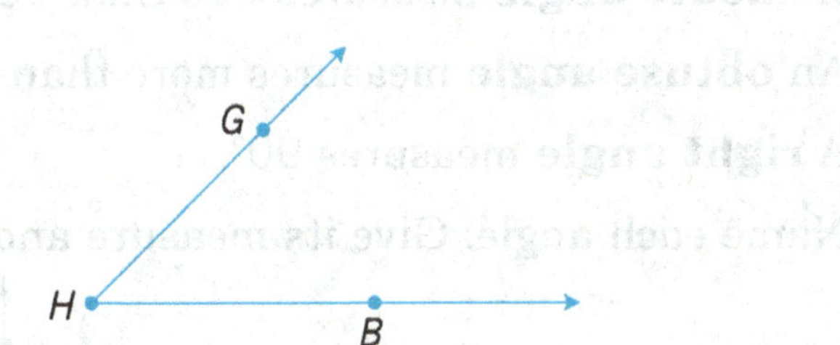

To use a protractor to measure an angle:

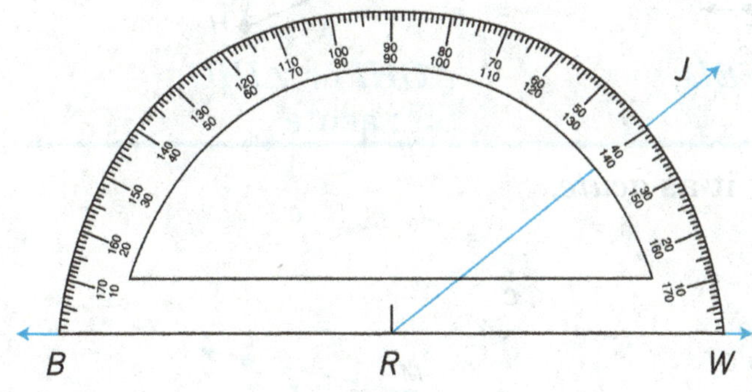

Place the center of the protractor at the vertex of the angle. Align one side of the angle with the base of the protractor. Use the scale starting at 0 and read the measure of the angle.

The measurement of ∠*JRW* is 40°.
The measurement of ∠*JRB* is 140°.

Name each angle. Then use a protractor to measure each angle.

	a	*b*	*c*
1.	∠____ ; ____°	∠____ ; ____°	∠____ ; ____°
2.	∠____ ; ____°	∠____ ; ____°	∠____ ; ____°

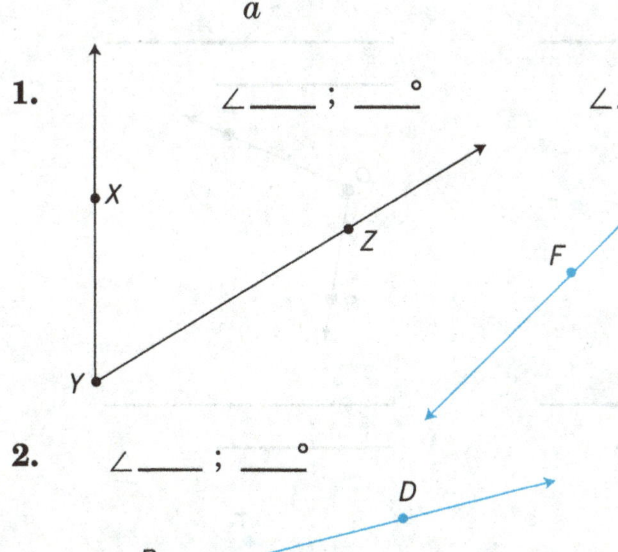

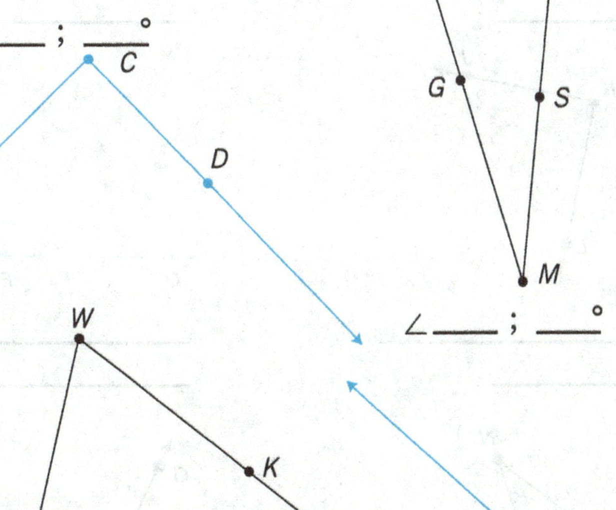

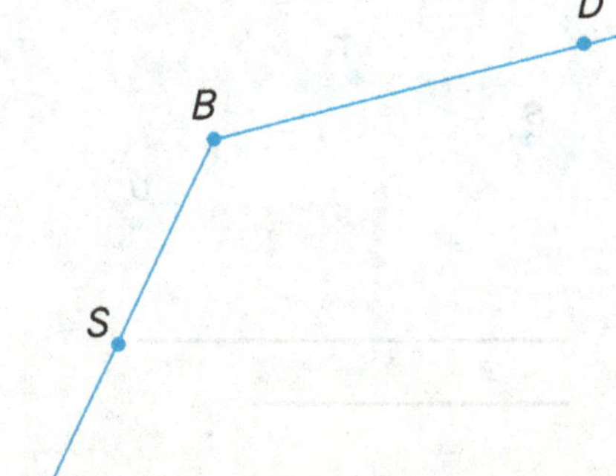

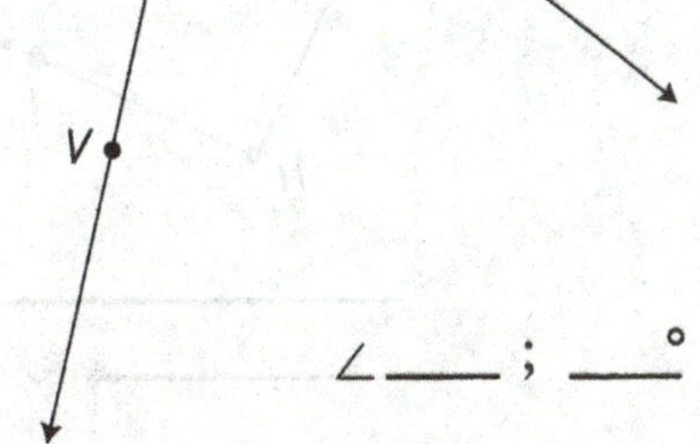

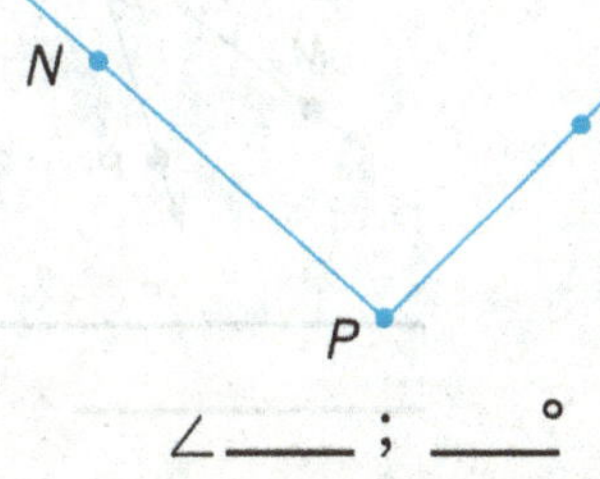

CHAPTER 14

Lesson 4 Acute, Obtuse, and Right Angles

An **acute angle** measures less than 90°.

An **obtuse angle** measures more than 90°.

A **right angle** measures 90°.

Name each angle. Give its measure and identify it as *acute, obtuse,* or *right.*

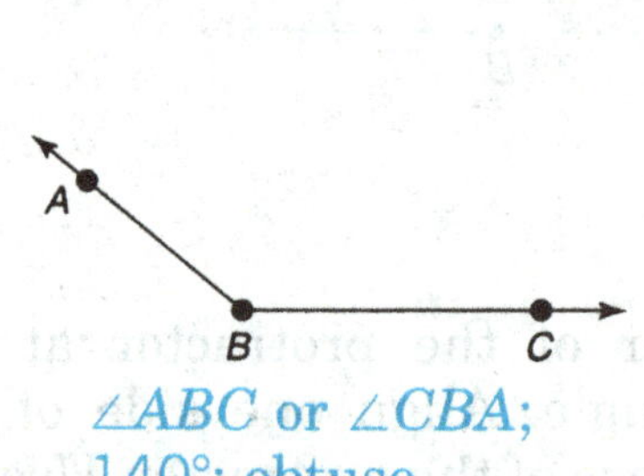

$\angle ABC$ or $\angle CBA$;
140°; obtuse

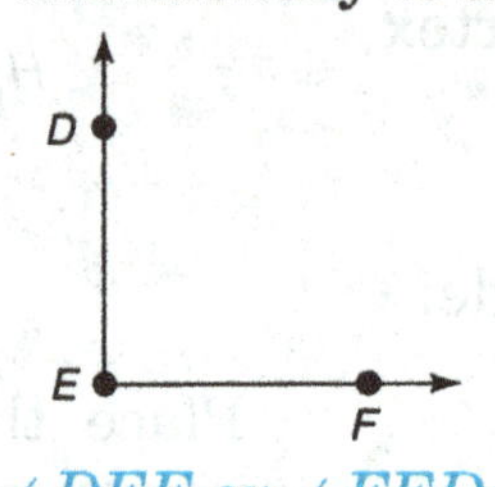

$\angle DEF$ or $\angle FED$;
90°; right

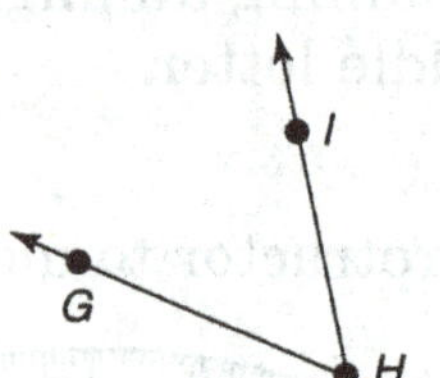

$\angle GHI$ or $\angle IHG$;
55°; acute

Name each angle. Give its measure and identify it as *acute, obtuse,* or *right.*

a *b* *c*

1.

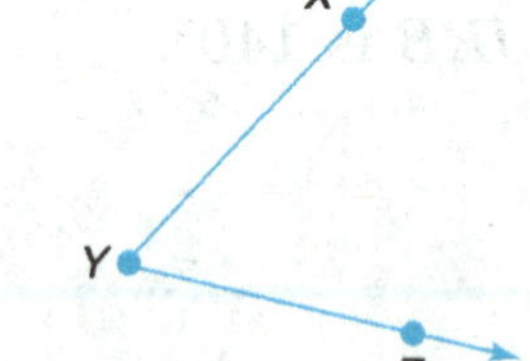

____
____

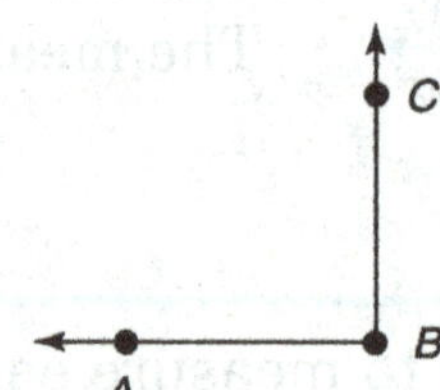

____
____

____
____

2.

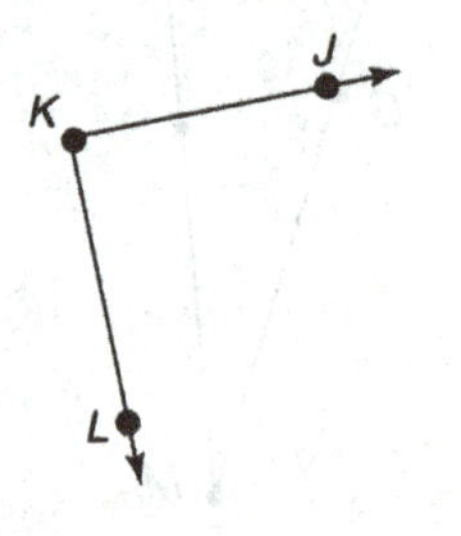

____
____

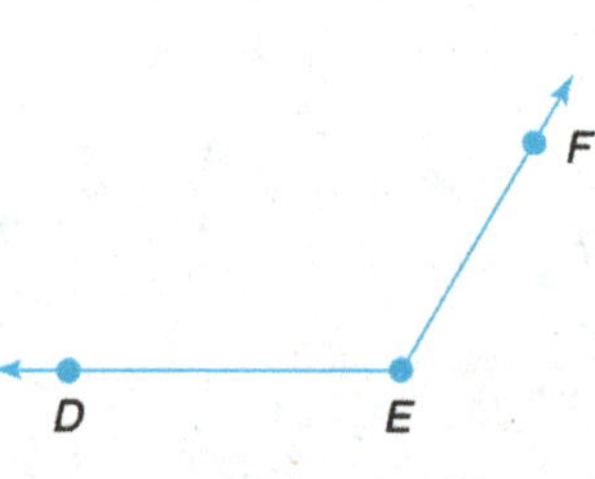

____
____

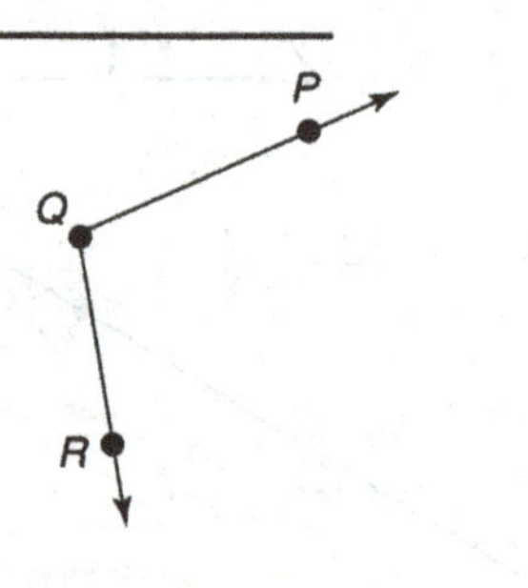

____
____

3.

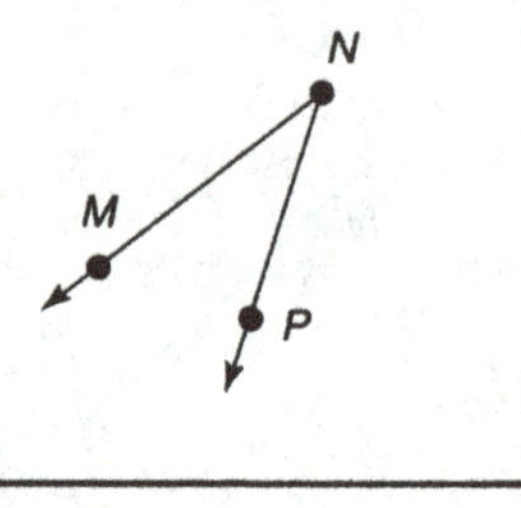

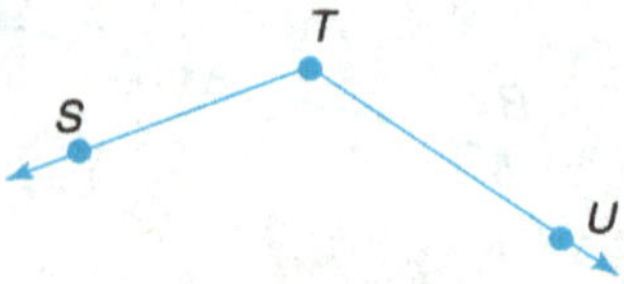

____
____

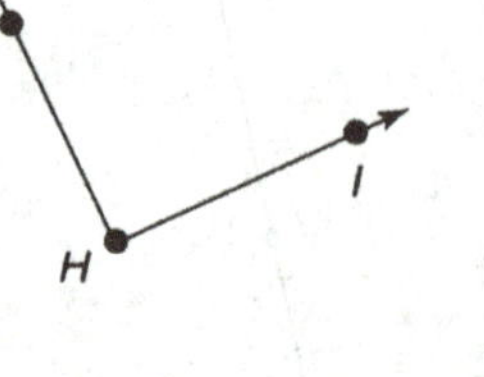

____
____

T
S
U

____
____

Lesson 5 Quadrilaterals

Quadrilaterals are figures that have 4 sides and 4 vertices. The following are more specific types of quadrilaterals.

A **parallelogram** has two pairs of opposite sides that are parallel.

A **rectangle** is a parallelogram with all right angles.

A **square** is a rectangle with all sides the same length.

A **trapezoid** has one pair of opposite sides that are parallel.

A **rhombus** is a parallelogram with all sides the same length.

Circle the name that best describes each quadrilateral.

1. square trapezoid parallelogram

2. rectangle square rhombus

3. rhombus trapezoid parallelogram

4.  square parallelogram rectangle

5. square rhombus rectangle

6. 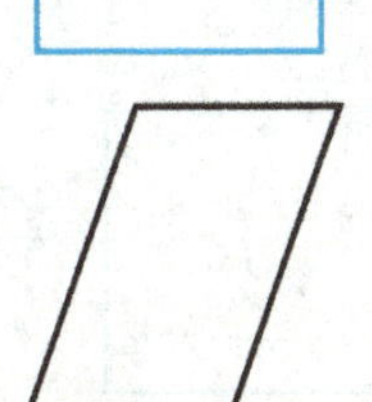 parallelogram trapezoid rectangle

NAME ____________________

Lesson 6 Polygons

A **polygon** is a closed shape that is formed by three or more sides. Polygons are named for the number of sides they have.

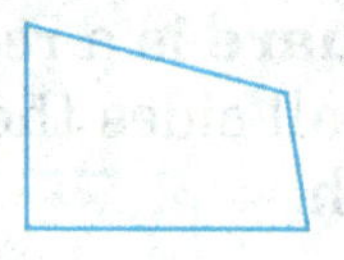

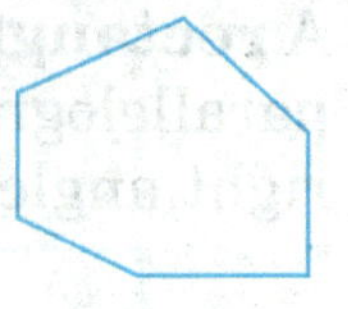
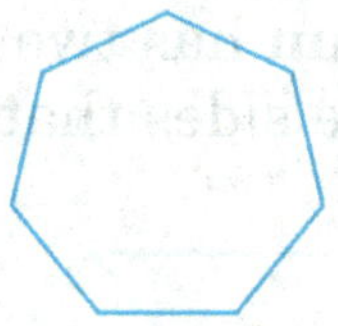

Triangle	**Quadrilateral**	**Pentagon**	**Hexagon**	**Heptagon**	**Octagon**
3 sides	___ sides	___ sides	___ sides	___ sides	___ sides

When all of the sides of a polygon are the same length, the figure is called a **regular polygon**.

The figure to the right is a regular hexagon.

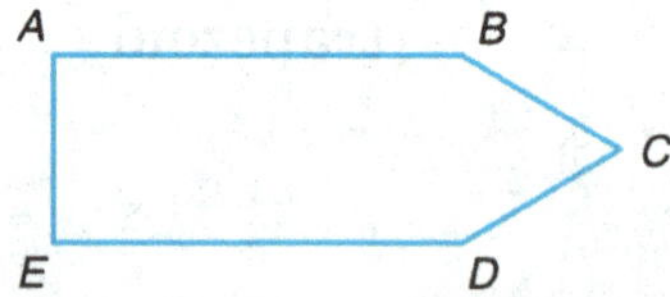

To name a polygon, use the letters of the vertices and list them in order.
The figure below is figure *ABCDE,* or pentagon *ABCDE.*

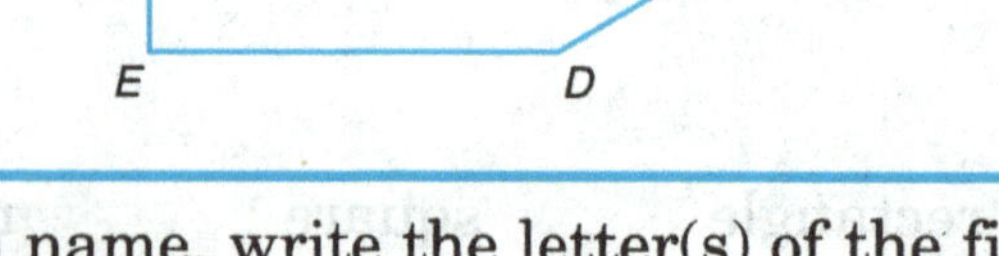

On the line after each name, write the letter(s) of the figure(s) it describes. Some names will have more than one letter. Some figures have more than one name.

1. pentagon ______
2. hexagon ______
3. octagon ______
4. triangle ______
5. heptagon ______
6. quadrilateral ______
7. regular triangle ______
8. regular hexagon ______
9. regular pentagon ______
10. regular quadrilateral ______

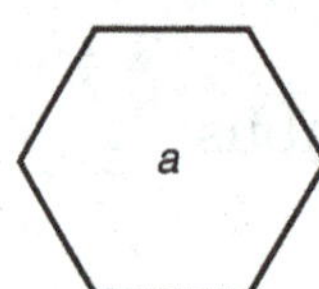

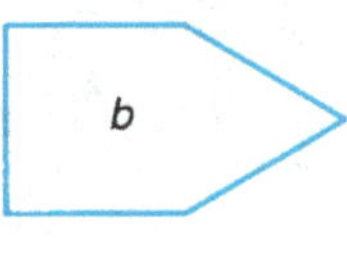

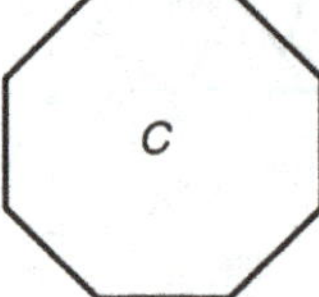

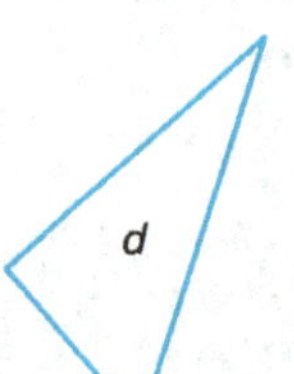

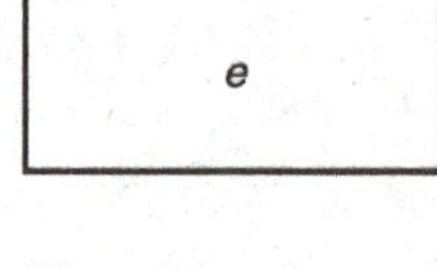

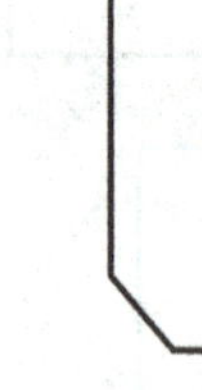

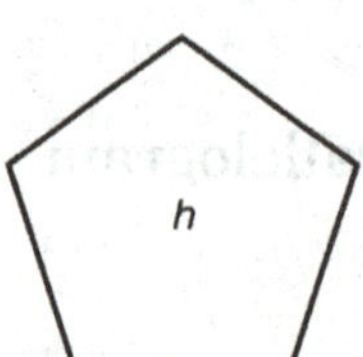

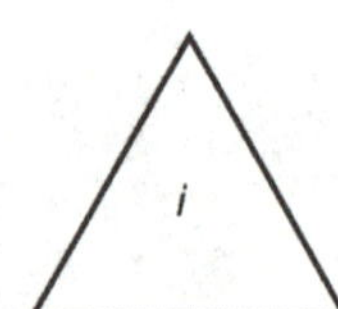

g

Lesson 6 Polygons

Name each figure shown.

a *b* *c*

11.

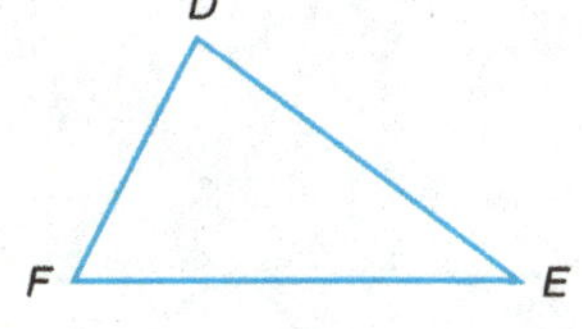

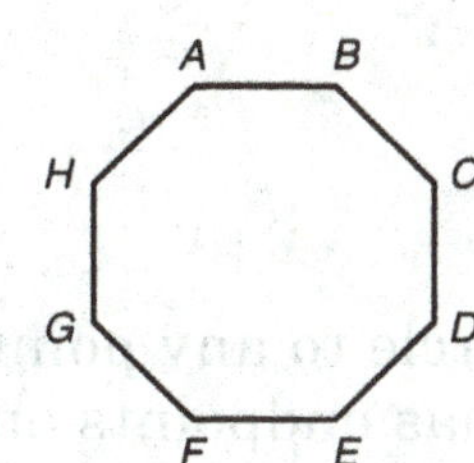

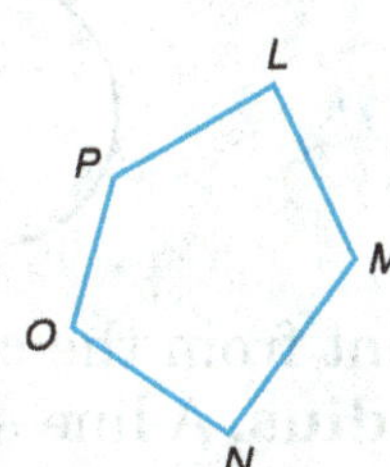

______________ ______________ ______________

12.

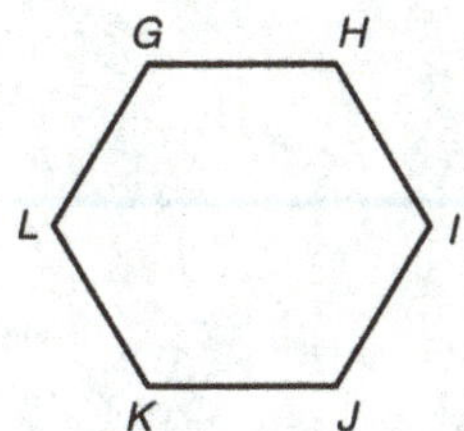

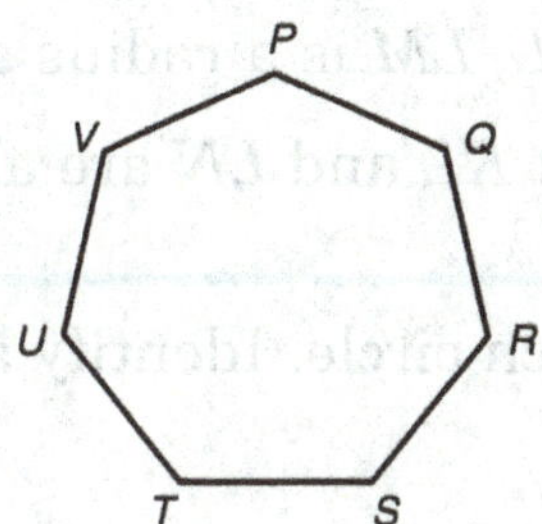

______________ ______________ ______________

A line segment that connects two vertices, but is not a side, is called a **diagonal.**

Two diagonals are drawn on pentagon *ABCDE*, diagonal *AD* and diagonal *EC*.

Draw and name all of the diagonals of each figure.

a *b*

13.

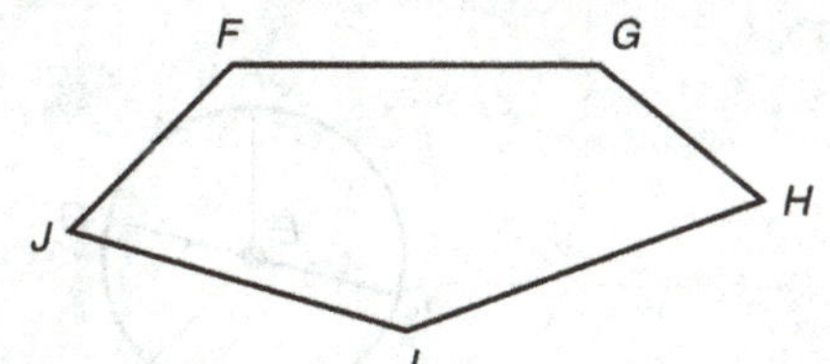

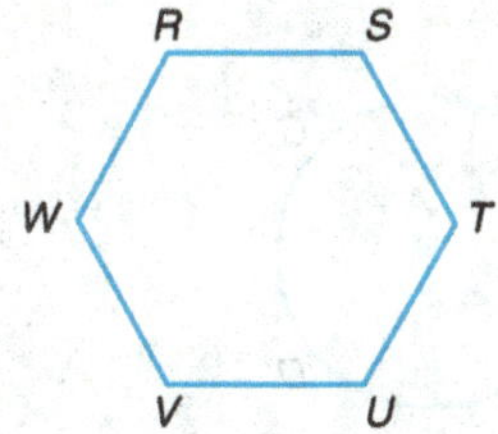

______________ ______________

Use problem **13** to answer the following questions.

14. Are all of the diagonals of figure *FGHIJ* the same length? ______

15. How many diagonals does figure *FGHIJ* have? ______

16. Are all of the diagonals of figure *RSTUVW* the same length? ______

17. How many diagonals does figure *RSTUVW* have? ______

NAME ____________________

Lesson 7 Circles

To name a **circle,** use the letter at the center of the circle.

Circle *P*

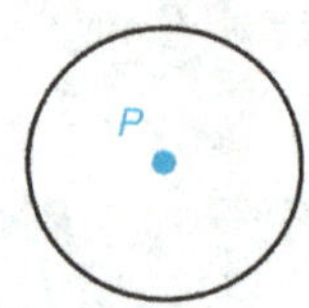

A line segment from the center of the circle to any point on the circle is a **radius.** A line segment that has endpoints on the circle and passes through the center of the circle is a **diameter.**

In circle *L*, *LM* is a radius and *KN* is a diameter.

Note that *KL* and *LN* are also **radii** (plural of radius).

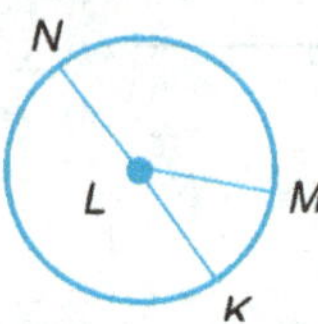

Name each circle. Identify a radius and diameter of each circle.

1.

a

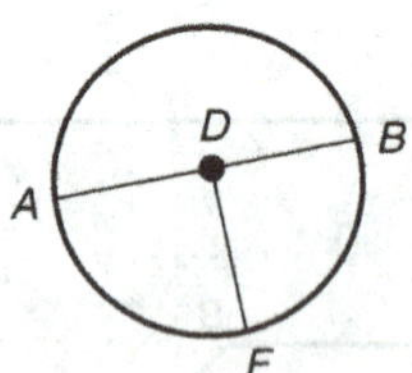

name: ____________

radius: ____________

____

diameter: ____________

b

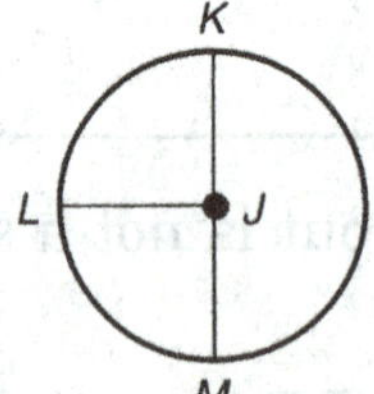

name: ____________

radius: ____________

____

diameter: ____________

c

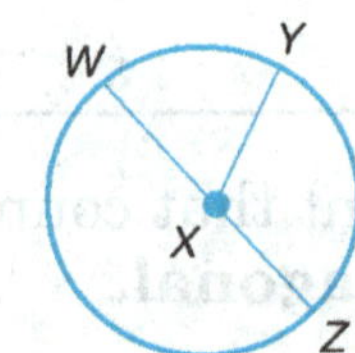

name: ____________

radius: ____________

____

diameter: ____________

2.

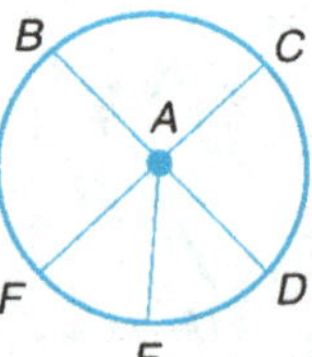

name: ____________

radius: ____________

____

diameter: ____________

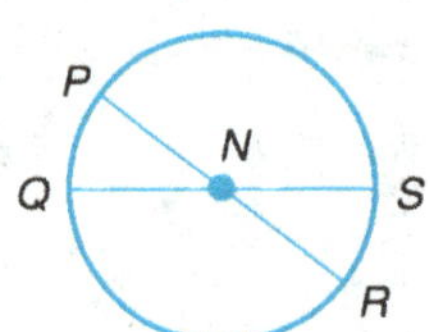

name: ____________

radius: ____________

____

diameter: ____________

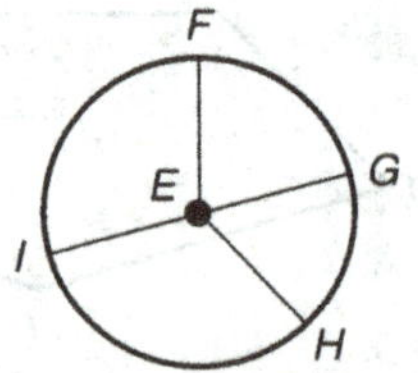

name: ____________

radius: ____________

diameter: ____________

3. Draw circle *S* with radius *ST* and diameter *QR*.

NAME ______________________

Lesson 8 Three-Dimensional Objects

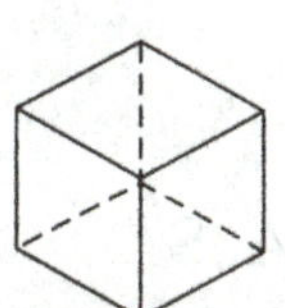
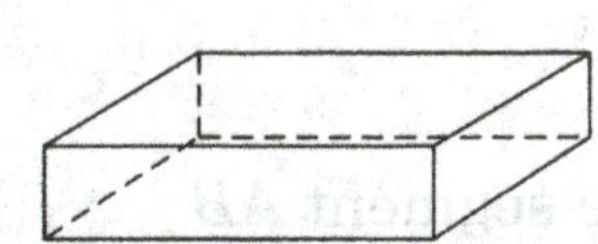
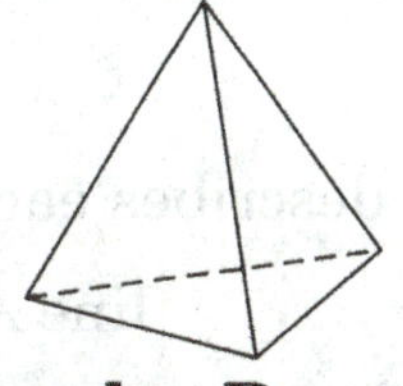
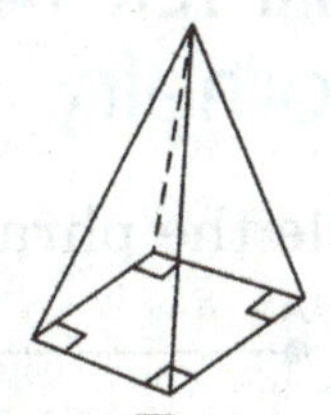

Cube **Rectangular Solid** **Triangular Pyramid** **Square Pyramid**

Each of these objects has faces, edges, and vertices.
Each of the faces of these objects is a polygon.

This is a **face**.

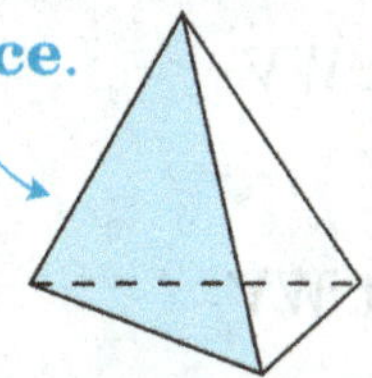

This is an **edge**.

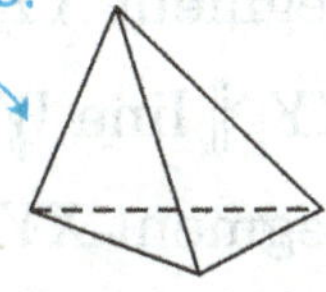

This is a **vertex**.

edge	face	rectangle	square	triangle	vertex
edges	faces	rectangles	squares	triangles	vertices

Choose from the list above to complete each sentence. You might use some words more than once. You might not use all the words.

1. All of the faces of a cube are ____________.

2. All of the faces of a rectangular solid are ____________.

3. The bottom face of a triangular pyramid is a ____________.

4. The colored part of object **A** below is a(n) ____________.

5. The colored part of object **B** below is a(n) ____________.

6. The colored part of object **C** below is a(n) ____________.

A 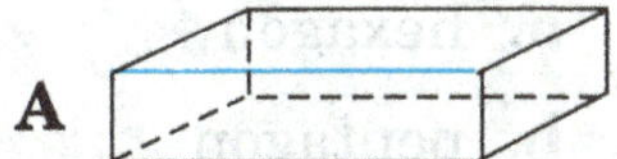

B 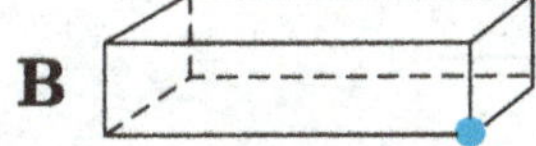

C 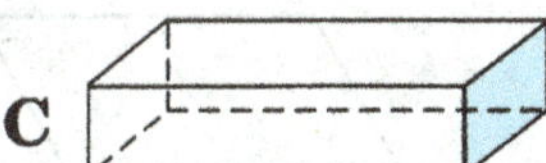

Answer each question with *Yes* or *No*.

7. Are all squares rectangles? __________

8. Are all faces of a cube rectangles? __________

9. Is a cube a rectangular solid? __________

CHAPTER 14

CHAPTER 14 PRACTICE TEST
Geometry

Circle the phrase that correctly describes each figure.

1. line AB line segment AB line A

2. line S line segment ST line ST

3. 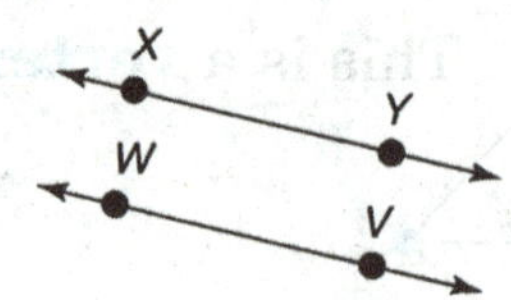

line segment $XY \parallel$ line segment WV

line $XY \parallel$ line WV

line segment $XY \perp$ line segment WV

4. line $FG \parallel$ line HJ

line $FG \perp$ line HJ

line segment $FG \perp$ line segment HJ

Name each angle. Give its measure and identify it as *acute, obtuse,* or *right.*

5. *a*

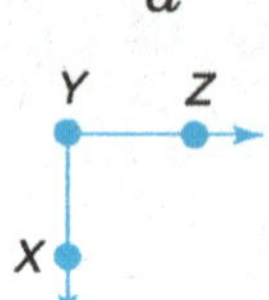

b

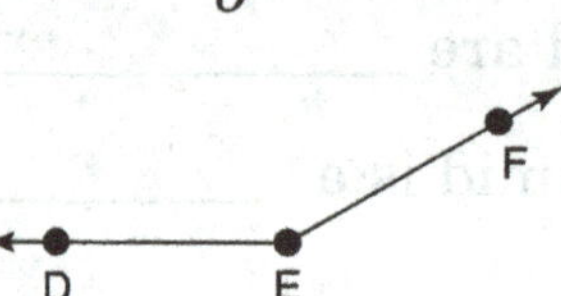

c

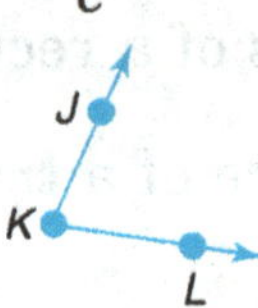

______________ ______________ ______________

______________ ______________ ______________

Write the letter for the name of each figure in the blank.

a *b*

6. ______ 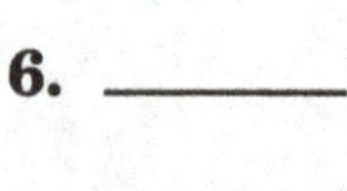 ______ 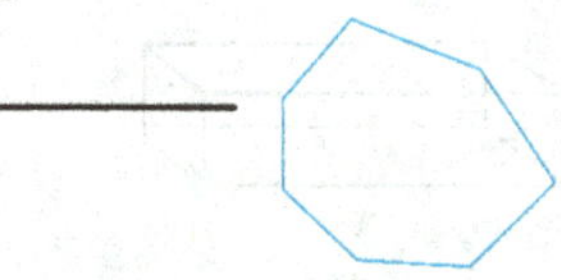

7. ______ ______

a. hexagon
b. pentagon
c. triangle
d. heptagon
e. circle
f. quadrilateral
g. octagon

NAME ______________________

MID-TEST Chapters 1-9

Solve each problem.

	a	*b*	*c*	*d*
1.	$\begin{array}{r} 42 \\ +51 \\ \hline \end{array}$	$\begin{array}{r} 92 \\ +78 \\ \hline \end{array}$	$\begin{array}{r} 134 \\ +939 \\ \hline \end{array}$	$\begin{array}{r} 46821 \\ 93289 \\ +25394 \\ \hline \end{array}$
2.	$\begin{array}{r} 75 \\ -18 \\ \hline \end{array}$	$\begin{array}{r} 236 \\ -57 \\ \hline \end{array}$	$\begin{array}{r} 1043 \\ -389 \\ \hline \end{array}$	$\begin{array}{r} 35670 \\ -34398 \\ \hline \end{array}$
3.	$\begin{array}{r} 78 \\ \times 5 \\ \hline \end{array}$	$\begin{array}{r} 147 \\ \times 9 \\ \hline \end{array}$	$\begin{array}{r} 850 \\ \times 8 \\ \hline \end{array}$	$\begin{array}{r} 3780 \\ \times 10 \\ \hline \end{array}$
4.	$\begin{array}{r} 37 \\ \times 28 \\ \hline \end{array}$	$\begin{array}{r} 92 \\ \times 40 \\ \hline \end{array}$	$\begin{array}{r} 248 \\ \times 75 \\ \hline \end{array}$	$\begin{array}{r} 1569 \\ \times 136 \\ \hline \end{array}$
5.	$6\overline{)96}$	$8\overline{)984}$	$9\overline{)3198}$	$73\overline{)7338}$

MID-TEST CH. 1-9

Solve each problem.

	a	b	c	d
6.	$5\overline{)92}$	$4\overline{)248}$	$49\overline{)1682}$	$89\overline{)17539}$
7.	$14\overline{)98}$	$9\overline{)186}$	$81\overline{)2734}$	$53\overline{)69791}$

Estimate the answer to each problem.

	a	b	c	d
8.	78 + 31	829 − 377	68 × 3	$5\overline{)241}$
9.	168 + 511	7294 − 867	684 × 17	$7\overline{)4367}$

Solve each problem.

	a	b	c	d
10.	\$3.75 + 9.12	\$5.18 − 3.41	\$4.29 × 5	$5\overline{)\$0.65}$
11.	\$291.06 + 88.64	\$714.62 − 126.33	\$11.63 × 24	$8\overline{)\$75.04}$

Use the bar graph to answer each question.

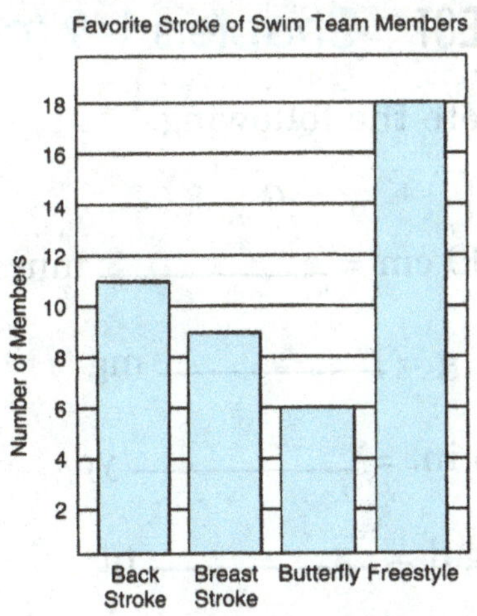

12. How many members chose butterfly as their favorite stroke? ____

13. How many members chose freestyle as their favorite stroke? ____

14. How many more members chose back stroke as their favorite stroke than breast stroke? ____

Use the line graph to answer each question.

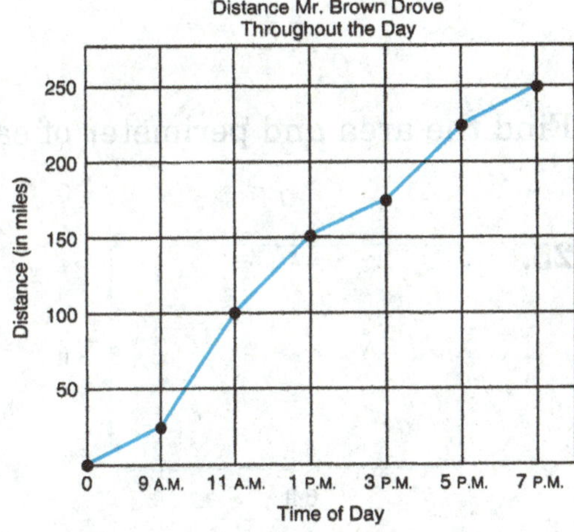

15. How many miles did Mr. Brown drive by 9:00 A.M.? ____

16. How many miles did Mr. Brown drive by 5:00 P.M.? ____

17. By what time had Mr. Brown driven 150 miles? ________

18. How many miles did Mr. Brown drive between 9:00 A.M. and 11:00 A.M.? ____

Find the mean, median, mode, and range of each set of numbers.

a

19. 2, 7, 4, 7, 5

mean: ____

median: ____

mode: ____

range: ____

b

33, 29, 45, 35, 36, 41, 33

mean: ____

median: ____

mode: ____

range: ____

20. 83%, 88%, 79%, 93%, 83%, 80%, 96%

mean: ____

median: ____

mode: ____

range: ____

$209, $218, $197, $224, $197

mean: ____

median: ____

mode: ____

range: ____

GO

MID-TEST CH. 1–9

Complete the following.

	a	*b*
21.	180 cm = ________ mm	3,000 liters = ________ kL
22.	21 g = ________ mg	300 kg = ________ g
23.	36 in. = ________ yd	1 mi 200 ft = ________ ft
24.	6 gal = ________ qt	8 lb 2 oz = ________ oz

Find the area and perimeter of each figure.

25. *a*

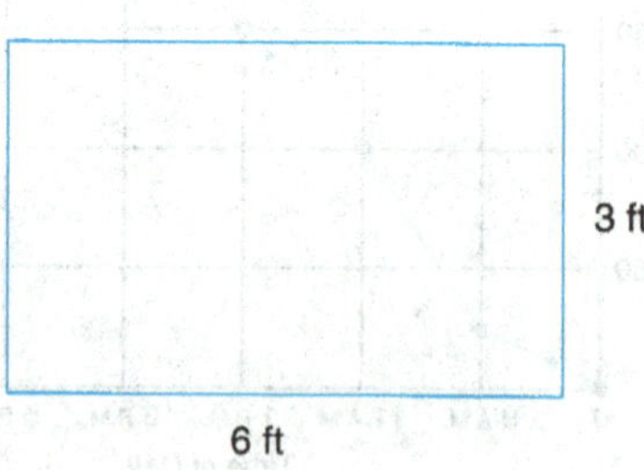

perimeter: ________ feet

area: ________ square feet

b

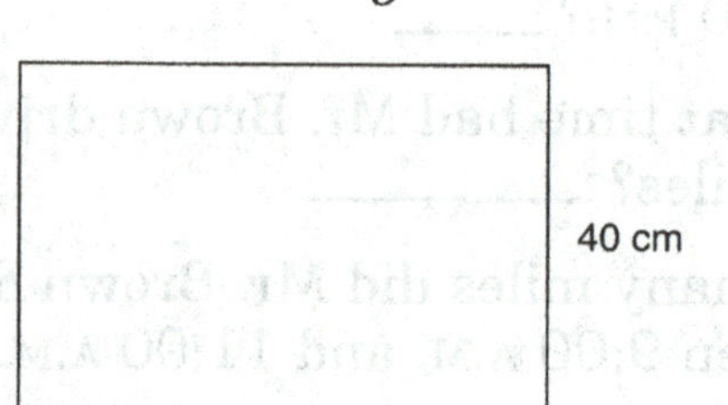

perimeter: ________ centimeters

area: ________ square centimeters

Find the volume of each figure.

26.

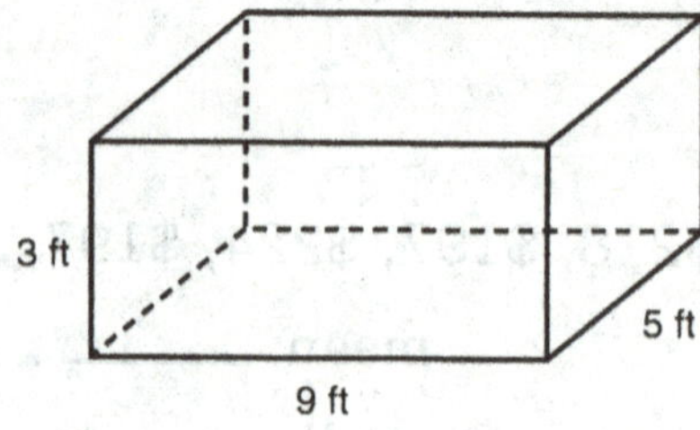

volume: ________ cubic feet

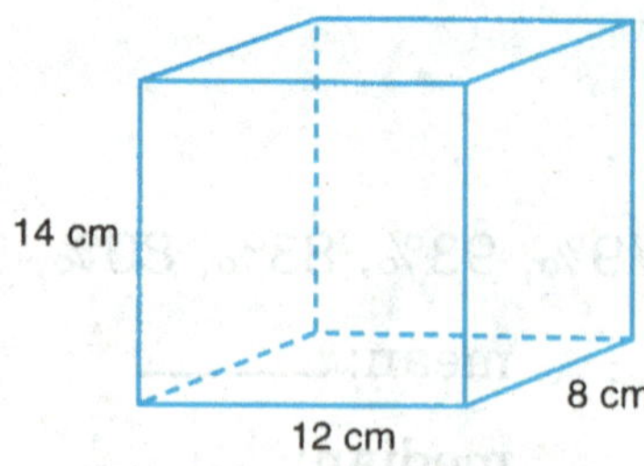

volume: ________ cubic centimeters

STOP

FINAL TEST Chapters 1–14

Solve each problem.

	a	*b*	*c*	*d*
1.	$\begin{array}{r} 36 \\ +57 \\ \hline \end{array}$	$\begin{array}{r} 798 \\ +135 \\ \hline \end{array}$	$\begin{array}{r} 45678 \\ +82902 \\ \hline \end{array}$	$\begin{array}{r} 7314 \\ 6452 \\ 9715 \\ +726 \\ \hline \end{array}$
2.	$\begin{array}{r} 63 \\ -18 \\ \hline \end{array}$	$\begin{array}{r} 178 \\ -65 \\ \hline \end{array}$	$\begin{array}{r} 1270 \\ -982 \\ \hline \end{array}$	$\begin{array}{r} 59246 \\ -37095 \\ \hline \end{array}$
3.	$\begin{array}{r} 73 \\ \times 6 \\ \hline \end{array}$	$\begin{array}{r} 124 \\ \times 8 \\ \hline \end{array}$	$\begin{array}{r} 387 \\ \times 10 \\ \hline \end{array}$	$\begin{array}{r} 420 \\ \times 32 \\ \hline \end{array}$
4.	$\begin{array}{r} 36 \\ \times 27 \\ \hline \end{array}$	$\begin{array}{r} 657 \\ \times 89 \\ \hline \end{array}$	$\begin{array}{r} 526 \\ \times 154 \\ \hline \end{array}$	$\begin{array}{r} 2984 \\ \times 697 \\ \hline \end{array}$
5.	$6\overline{)78}$	$89\overline{)10324}$	$9\overline{)3729}$	$51\overline{)6182}$
6.	$5\overline{)97}$	$4\overline{)231}$	$45\overline{)935}$	$93\overline{)27658}$

FINAL TEST CH. 1–14

Solve each problem.

	a	*b*	*c*	*d*
7.	$4.38 +1.07	$6.15 −3.82	$6.45 × 6	3)$0.81
8.	$379.25 +91.73	$843.26 −341.08	$21.18 × 32	7)$73.92

Find the mean, median, mode, and range of each set of numbers.

	a	*b*
9.	13, 18, 11, 18, 15	65, 58, 46, 53, 49, 61, 46
	mean: ________	mean: ________
	median: ________	median: ________
	mode: ________	mode: ________
	range: ________	range: ________

Complete the following.

	a	*b*
10.	160 cm = ________ mm	6 km = ________ m
11.	17 liters = ________ mL	9,000 liters = ________ kL
12.	9 g = ________ mg	30 kg = ________ g
13.	36 in. = ________ ft	1 mi 25 ft = ________ ft
14.	2 yd = ________ in.	2 qt 1 pt = ________ pt
15.	3 gal = ________ qt	6 lb 6 oz = ________ oz

GO

Find the area and perimeter of each figure.

16. *a*

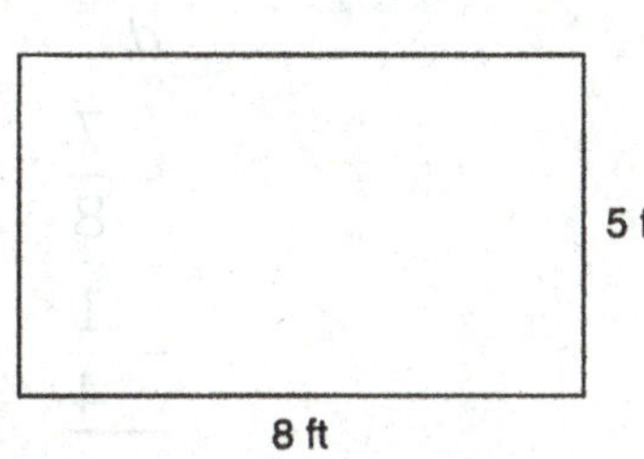

perimeter: ________ feet

area: ________ square feet

b

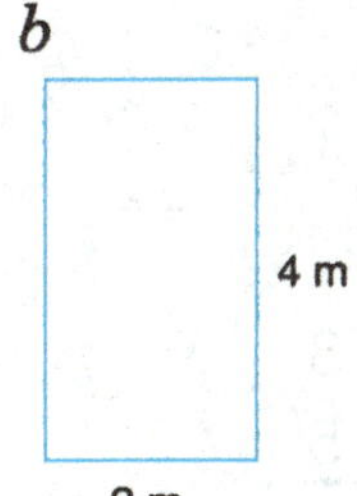

perimeter: ________ meters

area: ________ square meters

Change each fraction or mixed numeral to simplest form.

	a	*b*	*c*	*d*	*e*
17.	$\frac{9}{27} =$	$\frac{24}{30} =$	$\frac{35}{8} =$	$6\frac{4}{6} =$	$9\frac{16}{12} =$

Write each answer in simplest form.

	a	*b*	*c*
18.	$\frac{2}{3} \times \frac{1}{5} =$	$\frac{7}{8} \times \frac{1}{3} =$	$\frac{2}{7} \times \frac{3}{5} =$
19.	$1\frac{1}{3} \times \frac{2}{5} =$	$\frac{3}{4} \times 2\frac{3}{6} =$	$3 \times \frac{5}{6} =$
20.	$2\frac{1}{2} \times 3\frac{1}{3} =$	$\frac{7}{10} \times 5 =$	$4\frac{2}{5} \times 2\frac{3}{11} =$

	a	*b*	*c*	*d*
21.	$\frac{3}{5} + \frac{1}{5}$	$\frac{2}{7} + \frac{3}{7}$	$\frac{3}{10} + \frac{3}{10}$	$1\frac{3}{8} + 2\frac{1}{8}$
22.	$\frac{7}{8} + \frac{1}{4}$	$\frac{5}{12} + \frac{3}{4}$	$\frac{9}{10} + \frac{2}{3}$	$12\frac{3}{4} + 9\frac{2}{3}$

GO

FINAL TEST CH. 1–14

Write each answer in simplest form.

	a	*b*	*c*	*d*
23.	$\frac{7}{10} - \frac{3}{10}$	$7 - \frac{3}{5}$	$6\frac{3}{4} - 2\frac{1}{4}$	$\frac{7}{8} - \frac{1}{4}$
24.	$4\frac{7}{8} - 3\frac{2}{3}$	$6\frac{9}{10} - 1\frac{7}{8}$	$\frac{11}{12} - \frac{3}{4}$	$9\frac{1}{4} - 6\frac{2}{5}$

Name each figure.

25. ______________________

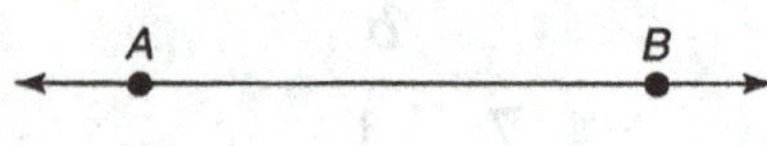

26. 

27. ______________________

28. ______________________

29. ______________________

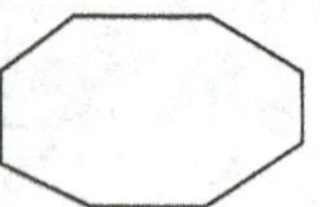

30. 

31. ______________________

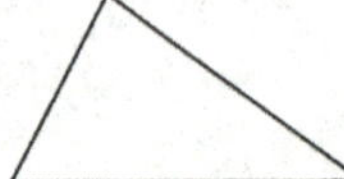

STOP

NAME ______________________

CHAPTER 1 CUMULATIVE REVIEW

Work each problem.
Find the correct answer.
Mark the space for the answer.

Part 1 Concepts

1. Which of the following addition problems has a sum closest to 500?

A 264 + 127
B 438 + 349
C 231 + 196
D 192 + 304

2. What is the estimated difference of 7819 − 4362?

F 3,000
G 4,000
H 5,000
J 12,000

3. What number is in the hundreds place in the sum of 6804 + 559?

A 3
B 5
C 6
D 7

4. Which of the following statements is true about the difference of 24,581 and 15,216?

F The difference is greater than 10,000.
G The difference is less than 9,000.
H The difference is an even number.
J The difference is an odd number.

Part 2 Computation

5. $\begin{array}{r} 92 \\ -67 \\ \hline \end{array}$

A 159 C 35
B 25 D 27

6. $\begin{array}{r} 718 \\ +425 \\ \hline \end{array}$

F 293 H 1,143
G 1,133 J 393

7. $\begin{array}{r} 6219 \\ -3468 \\ \hline \end{array}$

A 2,751
B 9,687
C 3,851
D 3,741

8. $\begin{array}{r} 8164 \\ +5457 \\ \hline \end{array}$

F 40,725
G 13,511
H 2,707
J 13,621

9. $\begin{array}{r} 149127 \\ -64853 \\ \hline \end{array}$

A 203,970
B 84,274
C 185,374
D 184,273

10. $\begin{array}{r} 41002 \\ -18653 \\ \hline \end{array}$

F 25,001
G 25,351
H 22,349
J 59,655

CUMULATIVE REVIEW

ANSWER ROW 1 Ⓐ Ⓑ Ⓒ Ⓓ 3 Ⓐ Ⓑ Ⓒ Ⓓ 5 Ⓐ Ⓑ Ⓒ Ⓓ 7 Ⓐ Ⓑ Ⓒ Ⓓ 9 Ⓐ Ⓑ Ⓒ Ⓓ
2 Ⓕ Ⓖ Ⓗ Ⓙ 4 Ⓕ Ⓖ Ⓗ Ⓙ 6 Ⓕ Ⓖ Ⓗ Ⓙ 8 Ⓕ Ⓖ Ⓗ Ⓙ 10 Ⓕ Ⓖ Ⓗ Ⓙ

11.

$$\begin{array}{r} 381 \\ 945 \\ +137 \\ \hline \end{array}$$

A 1,363
B 1,466
C 1,463
D 1,353

12.

$$\begin{array}{r} 3507 \\ 9124 \\ 7635 \\ +5180 \\ \hline \end{array}$$

F 24,336
G 25,446
H 26,346
J 15,447

Part 3 Applications

13. Casey rode his bike 18 miles on Saturday and 24 miles on Sunday. How many more miles did Casey ride his bike on Sunday?

A 42
B 7
C 16
D 6

14. On Monday, Joslyn read 24 pages of her book. On Tuesday, she read 37 pages of her book. On Wednesday, she read 33 pages of her book. How many total pages did Joslyn read on Monday, Tuesday, and Wednesday?

F 61
G 94
H 84
J 105

15. On Friday, Mr. Carmona drove 327 miles. On Saturday, he drove 189 miles. Use estimation to determine the approximate number of miles Mr. Carmona drove on Friday and Saturday.

A 300
B 600
C 500
D 400

16. Luisa and Jeff both flew to separate cities for business trips. Luisa's flight was 184 minutes long. Jeff's flight was 127 minutes long. How many minutes longer was Luisa's flight?

F 211
G 201
H 47
J 57

17. For a fund-raiser at school, students were selling candy bars. Amber sold 49 candy bars, Carlton sold 57 candy bars, and Takara sold 61 candy bars. Altogether, how many candy bars did Amber, Carlton, and Takara sell?

A 157
B 167
C 127
D 166

18. In 1990, the population of a small town was 16,843. In 2000, the population of the same town was 22,058. How much greater was the population in the year 2000 than in 1990?

F 5,215
G 38,891
H 38,901
J 5,415

STOP

ANSWER ROW 11 Ⓐ Ⓑ Ⓒ Ⓓ 13 Ⓐ Ⓑ Ⓒ Ⓓ 15 Ⓐ Ⓑ Ⓒ Ⓓ 17 Ⓐ Ⓑ Ⓒ Ⓓ
12 Ⓕ Ⓖ Ⓗ Ⓙ 14 Ⓕ Ⓖ Ⓗ Ⓙ 16 Ⓕ Ⓖ Ⓗ Ⓙ 18 Ⓕ Ⓖ Ⓗ Ⓙ

NAME ______________________

CHAPTER 2 CUMULATIVE REVIEW

Work each problem.
Find the correct answer.
Mark the space for the answer.

Part 1 Concepts

1. What number is in the tens place in the product of 34 and 43?

A 1 C 4
B 2 D 6

2. Which of the following products is the greatest?

F 34 × 9 H 49 × 5
G 18 × 15 J 68 × 4

3. Which problem would you use to find the estimated sum of 5,789 and 8,203?

A 6000 + 9000 C 5000 + 8000
B 6000 + 8000 D 5700 + 8200

4. What is the estimated product of 382 and 831?

F 240,000 H 320,000
G 360,000 J 270,000

5. What is the product of 15 and 153?

A 2,295 C 1,295
B 2,150 D 4,235

Part 2 Computation

6. $53 + 32$

F 21
G 75
H 32
J 85

7. $635 - 172$

A 807
B 463
C 707
D 563

8. $2834 - 368$

F 2,496
G 2,466
H 3,202
J 2,576

9. $8395 + 3705$

A 12,100
B 5,680
C 4,690
D 11,090

10. $869 + 432 + 257 + 118$

F 1,556
G 1,664
H 1,766
J 1,676

ANSWER ROW 1 Ⓐ Ⓑ Ⓒ Ⓓ 3 Ⓐ Ⓑ Ⓒ Ⓓ 5 Ⓐ Ⓑ Ⓒ Ⓓ 7 Ⓐ Ⓑ Ⓒ Ⓓ 9 Ⓐ Ⓑ Ⓒ Ⓓ
2 Ⓕ Ⓖ Ⓗ Ⓙ 4 Ⓕ Ⓖ Ⓗ Ⓙ 6 Ⓕ Ⓖ Ⓗ Ⓙ 8 Ⓕ Ⓖ Ⓗ Ⓙ 10 Ⓕ Ⓖ Ⓗ Ⓙ

CUMULATIVE REVIEW

11.

$$64 \times 5$$

A 320
B 119
C 69
D 300

12.

$$53 \times 21$$

F 814
G 1,103
H 1,113
J 159

13.

$$827 \times 60$$

A 48,220
B 15,757
C 50,220
D 49,620

14.

$$416 \times 209$$

F 85,894
G 86,944
H 4,576
J 86,953

Part 3 Applications

15. On a reading test, there were three sections. The number of points earned from each section is added together to get each test score. Marcy scored 38 points on the multiple choice section, 22 points on the matching section, and 31 points on the essay section. What was Marcy's test score?

A 81
B 90
C 91
D Not Given

16. Attendance at a high school basketball game for two consecutive games was 372 and 415. What was the difference in the attendance for these two games?

F 787
G 163
H 35
J 43

17. There are 6 buses taking students on a field trip to the zoo. Each bus is carrying 76 students. How many students are going on the field trip?

A 426
B 456
C 82
D 526

18. Emily's car can go 426 miles on one tank of gas. How many miles can her car go on 13 tanks of gas?

F 439
G 6,129
H 5,428
J Not Given

19. There are 34 rows of seats in the auditorium. Each row has 16 seats. How many seats are in the auditorium?

A 544
B 550
C 50
D 524

20. There are 168 hours in one week. How many hours are there in 52 weeks?

F 7,620
G 220
H 8,736
J 5,226

STOP

ANSWER ROW 11 Ⓐ Ⓑ Ⓒ Ⓓ 13 Ⓐ Ⓑ Ⓒ Ⓓ 15 Ⓐ Ⓑ Ⓒ Ⓓ 17 Ⓐ Ⓑ Ⓒ Ⓓ 19 Ⓐ Ⓑ Ⓒ Ⓓ
12 Ⓕ Ⓖ Ⓗ Ⓙ 14 Ⓕ Ⓖ Ⓗ Ⓙ 16 Ⓕ Ⓖ Ⓗ Ⓙ 18 Ⓕ Ⓖ Ⓗ Ⓙ 20 Ⓕ Ⓖ Ⓗ Ⓙ

NAME ______________________

CHAPTER 3 CUMULATIVE REVIEW

Work each problem.
Find the correct answer.
Mark the space for the answer.

Part 1 Concepts

1. Which of the following subtraction problems has a difference closest to 200?

   A 648 − 315 C 591 − 471
   B 918 − 725 D 784 − 507

2. What number is in the hundreds place in the sum of 6,734 and 4,318?

   F 0 H 1
   G 1 J 5

3. Which problem would you use to find the estimated product of 89 and 104?

   A 80 × 100 C 90 × 110
   B 90 × 100 D 80 × 110

4. Which of the following division problems has the greatest quotient?

   F $8\overline{)272}$ H $5\overline{)185}$
   G $4\overline{)132}$ J $9\overline{)288}$

5. Which problem would you use to find the estimated product of 593 and 287?

   A 600 × 300 C 590 × 280
   B 500 × 200 D 600 × 200

Part 2 Computation

6. 76 − 34

   F 100
   G 42
   H 110
   J 44

7. 614 + 237

   A 487
   B 841
   C 377
   D 851

8. 21763 − 9885

   F 11,878
   G 22,988
   H 31,648
   J 12,988

9. 819 × 6

   A 4,923
   B 1,485
   C 4,914
   D 4,864

10. 653 × 847

    F 12,407
    G 1,500
    H 544,191
    J 553,091

CUMULATIVE REVIEW

GO

ANSWER ROW 1 Ⓐ Ⓑ Ⓒ Ⓓ 3 Ⓐ Ⓑ Ⓒ Ⓓ 5 Ⓐ Ⓑ Ⓒ Ⓓ 7 Ⓐ Ⓑ Ⓒ Ⓓ 9 Ⓐ Ⓑ Ⓒ Ⓓ
2 Ⓕ Ⓖ Ⓗ Ⓙ 4 Ⓕ Ⓖ Ⓗ Ⓙ 6 Ⓕ Ⓖ Ⓗ Ⓙ 8 Ⓕ Ⓖ Ⓗ Ⓙ 10 Ⓕ Ⓖ Ⓗ Ⓙ

11. $4\overline{)72}$

A 17 r2 C 108
B 18 D 19

12. $8\overline{)658}$

F 82 r2 H 79 r7
G 81 J 93 r4

13. $9\overline{)2349}$

A 262 C 272 r1
B 334 r3 D 261

14. $7\overline{)95}$

F 10 r5 H 13 r4
G 13 J 14 r3

15. $5\overline{)648}$

A 132 C 125 r3
B 129 r3 D 29

Part 3 Applications

16. Kaya has 456 stickers in her sticker collection. Shanice has 507 stickers in her sticker collection. Together, how many stickers do Kaya and Shanice have in their sticker collections?

F 952 H 151
G 963 J 1,063

17. In one year, Mrs. Wessel drove her personal car 8,541 miles and she drove her business car 13,083 miles. How many more miles did Mrs. Wessel drive her company car?

A 5,543 C 21,624
B 4,632 D 4,542

18. In the soccer league, there are 8 teams with 17 members on each team. How many members are in the soccer league?

F 105 H 134
G 136 J 143

19. The grocery store received 23 boxes that contained cans of soup. Each box contains 48 cans of soup. How many cans of soup did the grocery store receive?

A 240 C 1,104
B 1,312 D 71

20. Each of the 23 students in Mr. Perez's class sold the same number of candy bars for a school fund-raiser. Mr. Perez's class sold a total of 621 candy bars. How many candy bars did each student in Mr. Perez's class sell?

F 22 H 23
G 25 J 27

ANSWER ROW 11 Ⓐ Ⓑ Ⓒ Ⓓ 13 Ⓐ Ⓑ Ⓒ Ⓓ 15 Ⓐ Ⓑ Ⓒ Ⓓ 17 Ⓐ Ⓑ Ⓒ Ⓓ 19 Ⓐ Ⓑ Ⓒ Ⓓ
12 Ⓕ Ⓖ Ⓗ Ⓙ 14 Ⓕ Ⓖ Ⓗ Ⓙ 16 Ⓕ Ⓖ Ⓗ Ⓙ 18 Ⓕ Ⓖ Ⓗ Ⓙ 20 Ⓕ Ⓖ Ⓗ Ⓙ

NAME ______________________

CHAPTER 4 CUMULATIVE REVIEW

Work each problem.
Find the correct answer.
Mark the space for the answer.

Part 1 Concepts

1. Which of the following products is the smallest?

A 168×23 C 93×45
B 71×64 D 108×36

2. What number is in the thousands place in the answer to the problem of 34081 + 18765?

F 5 H 2
G 4 J 8

3. What is the remainder of the division problem $8\overline{)371}$?

A 2 C 4
B 3 D 6

4. Which of the following division problems does not have a remainder?

F $15\overline{)943}$ H $37\overline{)6913}$
G $26\overline{)846}$ J $85\overline{)9435}$

Part 2 Computation

5. $\begin{array}{r} 4348 \\ +912 \\ \hline \end{array}$

A 4,436 C 5,260
B 4,250 D 3,436

6. $\begin{array}{r} 5906 \\ -2182 \\ \hline \end{array}$

F 3,724 H 3,824
G 7,088 J 8,088

7. $\begin{array}{r} 24 \\ \times 36 \\ \hline \end{array}$

A 660 C 60
B 7,344 D 864

8. $7\overline{)99}$

F 14 H 14 r1
G 12 J 12 r5

9. $5\overline{)805}$

A 141 C 159
B 161 D 181

10. $7\overline{)896}$

F 132 H 128
G 144 J 235

GO

CUMULATIVE REVIEW

ANSWER ROW 1 Ⓐ Ⓑ Ⓒ Ⓓ 3 Ⓐ Ⓑ Ⓒ Ⓓ 5 Ⓐ Ⓑ Ⓒ Ⓓ 7 Ⓐ Ⓑ Ⓒ Ⓓ 9 Ⓐ Ⓑ Ⓒ Ⓓ
2 Ⓕ Ⓖ Ⓗ Ⓙ 4 Ⓕ Ⓖ Ⓗ Ⓙ 6 Ⓕ Ⓖ Ⓗ Ⓙ 8 Ⓕ Ⓖ Ⓗ Ⓙ 10 Ⓕ Ⓖ Ⓗ Ⓙ

11. $15\overline{)621}$

A 41 r6
B 41
C 56 r11
D 41 r5

12. $27\overline{)729}$

F 34 r11
G 27 r20
H 27
J 21 r19

13. $71\overline{)1474}$

A 30 r44
B 20 r54
C 19 r25
D 21 r8

Part 3 Applications

14. There are 148 pages in Antwan's new book. He has read 85 pages. How many pages does he have left to read?

F 68
G 63
H 70
J 85

15. Attendance at a college football stadium for four consecutive games was 72,369; 69,516; 67,842; and 71,096. What was the total attendance for all four games?

A 209,727
B 280,823
C 279,422
D 380,823

16. Each box of cereal weighs 15 ounces. What is the total weight of 8 boxes?

F 150 ounces
G 23 ounces
H 140 ounces
J 120 ounces

17. The airline distance between two cities is 462 miles. How many miles would an airplane travel in making 34 one-way trips?

A 14,708
B 15,608
C 15,708
D 15,508

18. There are 115 students in the fifth grade at Middletown Elementary school. There are five different classrooms with the same number of students in each. How many students are in each classroom?

F 23
G 25
H 29
J 21

19. Six students shared the cost of a weekend camping trip. If the total cost of the trip was $348, what was each student's share?

A $116
B $58
C $48
D $34

20. A machine operated for 48 hours and produced 8,016 parts. The same number of parts were produced each hour. How many parts were produced each hour?

F 4,008
G 248
H 159
J 167

STOP

ANSWER ROW 11 Ⓐ Ⓑ Ⓒ Ⓓ 13 Ⓐ Ⓑ Ⓒ Ⓓ 15 Ⓐ Ⓑ Ⓒ Ⓓ 17 Ⓐ Ⓑ Ⓒ Ⓓ 19 Ⓐ Ⓑ Ⓒ Ⓓ
12 Ⓕ Ⓖ Ⓗ Ⓙ 14 Ⓕ Ⓖ Ⓗ Ⓙ 16 Ⓕ Ⓖ Ⓗ Ⓙ 18 Ⓕ Ⓖ Ⓗ Ⓙ 20 Ⓕ Ⓖ Ⓗ Ⓙ

CHAPTER 5 CUMULATIVE REVIEW

NAME ______________________

Work each problem.
Find the correct answer.
Mark the space for the answer.

Part 1 Concepts

1. What is the sum of eighty-three and fifty-nine?

A 132 C 112
B 4,897 D 142

2. What digit is in the hundreds place in the difference of 4,703 and 2,576?

F 3 H 2
G 0 J 1

3. What is the remainder of the division problem $21\overline{)947}$?

A 2 C 4
B 9 D 21

4. What number is the dividend in the problem $85\overline{)5307}$?

F 62 H 85
G 5,307 J 36

5. What number is the divisor in the problem $46\overline{)989}$?

A 989 C 20
B 21 D 46

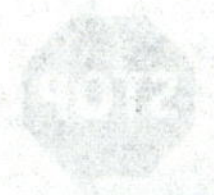

Part 2 Computation

6.
$$\begin{array}{r} 36 \\ +57 \\ \hline \end{array}$$

F 83
G 93
H 94
J 103

7.
$$\begin{array}{r} 17362 \\ -9548 \\ \hline \end{array}$$

A 7,814
B 8,814
C 7,824
D 7,816

8.
$$\begin{array}{r} 45 \\ \times 16 \\ \hline \end{array}$$

F 690
G 720
H 29
J 72

9. $6\overline{)84}$

A 24
B 504
C 18
D 14

10. $5\overline{)4106}$

F 821 r6
G 802 r1
H 821 r1
J 820 r6

CUMULATIVE REVIEW

ANSWER ROW 1 Ⓐ Ⓑ Ⓒ Ⓓ 3 Ⓐ Ⓑ Ⓒ Ⓓ 5 Ⓐ Ⓑ Ⓒ Ⓓ 7 Ⓐ Ⓑ Ⓒ Ⓓ 9 Ⓐ Ⓑ Ⓒ Ⓓ
2 Ⓕ Ⓖ Ⓗ Ⓙ 4 Ⓕ Ⓖ Ⓗ Ⓙ 6 Ⓕ Ⓖ Ⓗ Ⓙ 8 Ⓕ Ⓖ Ⓗ Ⓙ 10 Ⓕ Ⓖ Ⓗ Ⓙ

11. $12\overline{)3000}$

A 250
B 25
C 205
D 600

12. $34\overline{)850}$

F 25 r20
G 25
H 27 r32
J 816

13. $73\overline{)4161}$

A 4,088
B 67
C 57
D 61

14. $47\overline{)35918}$

F 762 r4
G 753
H 764 r4
J 764 r10

Part 3 Applications

15. The number of parts shipped to three different countries was 671, 818, and 448. How many parts were shipped in all?

A 1,927 C 1,937
B 1,837 D 1,827

16. It takes Samantha 18 minutes to run around the lake. How many minutes will it take Samantha to run around the lake four times?

F 22 H 60
G 54 J 72

17. A factory shipped 858 tires to six different cities. Each city received the same number of tires. How many tires did each city receive?

A 142 C 134
B 143 D 192

18. In Section C of an arena, there are 450 seats. Each row has 18 seats. How many rows are in Section C?

F 25 H 35
G 8,100 J 7,100

19. There are 2,941 potatoes to be put into bags. Each bag will contain 17 potatoes. How many bags will be needed?

A 2,924 C 17
B 183 D 173

20. During 3 months, 93 employees worked 42,408 hours. Each employee worked the same number of hours. How many hours did each employee work during the 3 months?

F 456 H 14,136
G 279 J 269

STOP

ANSWER ROW 11 Ⓐ Ⓑ Ⓒ Ⓓ 13 Ⓐ Ⓑ Ⓒ Ⓓ 15 Ⓐ Ⓑ Ⓒ Ⓓ 17 Ⓐ Ⓑ Ⓒ Ⓓ 19 Ⓐ Ⓑ Ⓒ Ⓓ
12 Ⓕ Ⓖ Ⓗ Ⓙ 14 Ⓕ Ⓖ Ⓗ Ⓙ 16 Ⓕ Ⓖ Ⓗ Ⓙ 18 Ⓕ Ⓖ Ⓗ Ⓙ 20 Ⓕ Ⓖ Ⓗ Ⓙ

NAME ____________________

CHAPTER 6 CUMULATIVE REVIEW

Work each problem.
Find the correct answer.
Mark the space for the answer.

Part 1 Concepts

1. Which of the following products is the greatest?

A 46×49 C 33×65
B 29×87 D 72×18

2. What number is the divisor in the problem $34\overline{)7438}$?

F 7,438 H 34
G 218 J 26

3. What number is in the hundredths place in $1,529.74?

A 5 C 7
B 2 D 4

4. Write $924.08 using words.

F nine twenty-four and eight tenths
G nine hundred twenty-four dollars and eighty cents
H nine hundred twenty-four dollars and eight cents
J ninety-two dollars and eight cents

5. What number is in the ones place in $538.29?

A 8 C 5
B 9 D 3

Part 2 Computation

6. $798 + 135$

F 923
G 933
H 833
J 823

7. $1270 - 982$

A 2,252
B 298
C 398
D 288

8. 614×53

F 32,542
G 4,912
H 31,542
J 30,542

9. $3\overline{)741}$

A 271
B 247
C 410
D 120

10. $63\overline{)2923}$

F 40 r3
G 46 r35
H 46 r63
J 46 r25

CUMULATIVE REVIEW

ANSWER ROW 1 Ⓐ Ⓑ Ⓒ Ⓓ 3 Ⓐ Ⓑ Ⓒ Ⓓ 5 Ⓐ Ⓑ Ⓒ Ⓓ 7 Ⓐ Ⓑ Ⓒ Ⓓ 9 Ⓐ Ⓑ Ⓒ Ⓓ
2 Ⓕ Ⓖ Ⓗ Ⓙ 4 Ⓕ Ⓖ Ⓗ Ⓙ 6 Ⓕ Ⓖ Ⓗ Ⓙ 8 Ⓕ Ⓖ Ⓗ Ⓙ 10 Ⓕ Ⓖ Ⓗ Ⓙ

11. 15)1470

A 90 r8
B 90 r12
C 98
D 90

12.

$84.23
−61.94

F $23.29
G $22.29
H $146.17
J $157.18

13.

$52.25
× 6

A $313.20
B $3,135.00
C $118.91
D $313.50

14. 8)$11.92

F $1.49
G $1.47
H $1.39
J $1.52

Part 3 Applications

15. Mr. Allen had 86,963 miles on his car at the end of January. By the end of February, he had 88,042 miles on his car. How many miles did Mr. Allen drive his car in February?

A 2,198 C 1,079
B 88,042 D 1,457

16. A truck driver is delivering boxes of books to schools. Each box contains 12 books. He delivers 25 boxes to one school. How many books did he deliver to that school?

F 300 H 275
G 12 J 13

17. There are 93 middle school classrooms in the Plainsville School District. These classrooms hold 2,604 students. Each classroom has the same number of students. How many students are there in each classroom?

A 30 C 32
B 28 D 23

18. A carton holds 24 cans of juice. How many cartons will it take to hold 1,512 cans?

F 60 H 71
G 63 J 80

19. Mrs. Anderson spent $73.95 at the grocery store. She gave the cashier $80.00. How much change did Mrs. Anderson get back?

A $7.15 C $6.05
B $153.95 D $5.55

20. Ben and four of his friends ordered pizzas. The bill came to $19.00. They split the bill evenly among all of them. How much did each boy pay?

F $3.80 H $95.00
G $4.75 J $3.17

STOP

ANSWER ROW 11 Ⓐ Ⓑ Ⓒ Ⓓ 13 Ⓐ Ⓑ Ⓒ Ⓓ 15 Ⓐ Ⓑ Ⓒ Ⓓ 17 Ⓐ Ⓑ Ⓒ Ⓓ 19 Ⓐ Ⓑ Ⓒ Ⓓ
12 Ⓕ Ⓖ Ⓗ Ⓙ 14 Ⓕ Ⓖ Ⓗ Ⓙ 16 Ⓕ Ⓖ Ⓗ Ⓙ 18 Ⓕ Ⓖ Ⓗ Ⓙ 20 Ⓕ Ⓖ Ⓗ Ⓙ

NAME ______________________

CHAPTER 7 CUMULATIVE REVIEW

Work each problem.
Find the correct answer.
Mark the space for the answer.

Part 1 Concepts

1. Which problem would you use to find the estimated difference of 831 and 483?

A 800 − 400 C 900 − 400
B 800 − 500 D 900 − 500

2. Which of the following dollar amounts has a 4 in the tenths place?

F \$926.45 H \$308.64
G \$754.91 J \$145.07

3. Which of the following words means "average"?

A median C range
B mode D mean

4. Which of the following means the number that appears most often in a set of numbers?

F median H range
G mode J mean

5. Which of the following words means the chance that something will occur?

A mode C median
B range D probability

Part 2 Computation

6. 1356 + 793 + 2684

F 3,833
G 4,833
H 4,823
J 4,723

7. 754 × 32

A 24,028
B 23,128
C 24,128
D 23,028

8. 6)3892

F 640 r5
G 644 r4
H 650 r5
J 648 r4

9. 45)3150

A 7
B 31
C 70
D 75

10. 61)7285

F 119 r26
G 119 r42
H 119 r34
J 119 r43

CUMULATIVE REVIEW

ANSWER ROW 1 Ⓐ Ⓑ Ⓒ Ⓓ 3 Ⓐ Ⓑ Ⓒ Ⓓ 5 Ⓐ Ⓑ Ⓒ Ⓓ 7 Ⓐ Ⓑ Ⓒ Ⓓ 9 Ⓐ Ⓑ Ⓒ Ⓓ
2 Ⓕ Ⓖ Ⓗ Ⓙ 4 Ⓕ Ⓖ Ⓗ Ⓙ 6 Ⓕ Ⓖ Ⓗ Ⓙ 8 Ⓕ Ⓖ Ⓗ Ⓙ 10 Ⓕ Ⓖ Ⓗ Ⓙ

11. $\begin{array}{r} \$452.35 \\ +986.54 \\ \hline \end{array}$

A $1,506.65
C $1,439.89
B $1,438.89
D $1,338.90

12. $\begin{array}{r} \$913.55 \\ -639.12 \\ \hline \end{array}$

F $374.43
H $274.43
G $293.42
J $294.43

13. $\begin{array}{r} \$3.47 \\ \times 12 \\ \hline \end{array}$

A $41.74
C $41.77
B $40.64
D $41.64

14. What is the mean of 11, 16, 9, 13, and 11?

F 7
H 11
G 12
J 10

15. A number cube has sides labeled A, B, C, D, E, and F. What is the probability of rolling an E?

A $\frac{1}{3}$
C $\frac{1}{6}$
B $\frac{1}{2}$
D $\frac{1}{5}$

Part 3 Applications

16. On Monday, Shontay drove 293 miles. On Tuesday, she drove 137 miles. How many more miles did Shontay drive on Monday than on Tuesday?

F 430
H 137
G 146
J 156

17. Maggie bought 6 T-shirts. Each T-shirt cost $13.75. How much money did Maggie spend on the T-shirts?

A $82.50
C $68.50
B $78.50
D $78.20

18. There are 894 boxes ready to be shipped. Only 38 boxes will fit in each truck. How many trucks will be needed to ship all the boxes?

F 23
H 38
G 20
J 24

19. Kenji, Sarah, Felix, and Darnell spent 3 hours on Saturday raking leaves. They were paid $22.00. They split the money evenly among them. How much money did each receive?

A $7.34
C $5.50
B $12.00
D $7.50

Use this graph to answer question **20.**

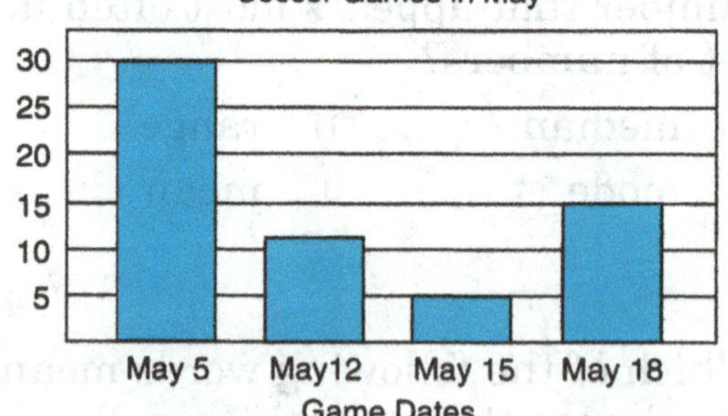

20. How many more fans attended the May 5 game than the May 15 game?

F 15
H 19
G 25
J 23

STOP

ANSWER ROW 11 Ⓐ Ⓑ Ⓒ Ⓓ 13 Ⓐ Ⓑ Ⓒ Ⓓ 15 Ⓐ Ⓑ Ⓒ Ⓓ 17 Ⓐ Ⓑ Ⓒ Ⓓ 19 Ⓐ Ⓑ Ⓒ Ⓓ
12 Ⓕ Ⓖ Ⓗ Ⓙ 14 Ⓕ Ⓖ Ⓗ Ⓙ 16 Ⓕ Ⓖ Ⓗ Ⓙ 18 Ⓕ Ⓖ Ⓗ Ⓙ 20 Ⓕ Ⓖ Ⓗ Ⓙ

NAME ______________

CHAPTER 8 CUMULATIVE REVIEW

Work each problem.
Find the correct answer.
Mark the space for the answer.

Part 1 Concepts

1. What digit is in the ones place in the quotient of 14)2567?

A 4   C 3
B 7   D 5

2. Which of the following sums is the greatest?

F $839.73 + 274.11
G $438.27 + 728.10
H $910.35 + 183.49
J $602.76 + 538.27

3. Which of the following means the difference between the greatest and least number in a set of numbers?

A mean   C mode
B median   D range

4. Which of the following units of measure could represent the length of a room?

F 13 liters   H 5 meters
G 13 kilograms   J 5 grams

5. Which of the following products is the smallest?

A 65 × 14   C 114 × 10
B 181 × 8   D 88 × 57

Part 2 Computation

6.

$$\begin{array}{r} 3067 \\ -1372 \\ \hline \end{array}$$

F 1,745
G 1,695
H 2,695
J 1,795

7.

$$\begin{array}{r} 354 \\ \times 7 \\ \hline \end{array}$$

A 2,458
B 2,178
C 2,478
D 2,358

8. 8)89

F 10 r1
G 11 r1
H 9 r8
J 8 r1

9. 24)$39.12

A $1.63
B $1.12
C $1.36
D $1.53

10.

$$\begin{array}{r} \$839.46 \\ +173.45 \\ \hline \end{array}$$

F $1,042.91
G $942.81
H $1,042.81
J $1,012.91

CUMULATIVE REVIEW

ANSWER ROW 1 Ⓐ Ⓑ Ⓒ Ⓓ 3 Ⓐ Ⓑ Ⓒ Ⓓ 5 Ⓐ Ⓑ Ⓒ Ⓓ 7 Ⓐ Ⓑ Ⓒ Ⓓ 9 Ⓐ Ⓑ Ⓒ Ⓓ
2 Ⓕ Ⓖ Ⓗ Ⓙ 4 Ⓕ Ⓖ Ⓗ Ⓙ 6 Ⓕ Ⓖ Ⓗ Ⓙ 8 Ⓕ Ⓖ Ⓗ Ⓙ 10 Ⓕ Ⓖ Ⓗ Ⓙ

11. What is the mode of 46, 38, 36, 31, 47, 32, and 36?

A 31 C 38
B 36 D 16

12. 6 cm = ________ mm

F 6 H 60
G 600 J 6,000

13. 5,000 mg = ________ grams

A 5 C 50
B 500 D 5,000,000

14. What is the perimeter of the figure shown?

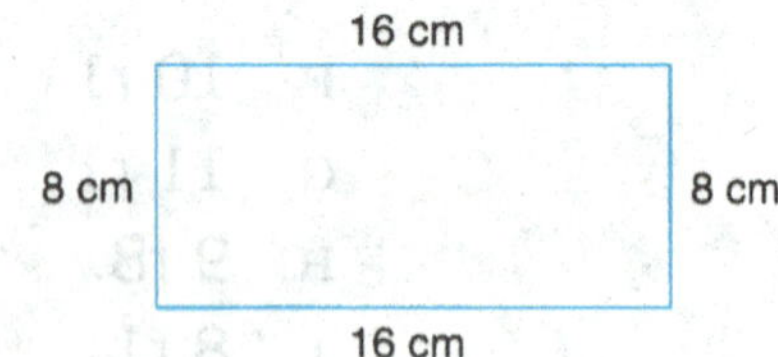

F 40 cm H 128 cm
G 48 cm J 64 cm

Part 3 Applications

15. Marcus rode his bike four days this week. Monday he biked 15 miles. Wednesday he biked 9 miles. Thursday he biked 13 miles. Saturday he biked 21 miles. How many miles did Marcus bike this week?

A 50 C 58
B 48 D 40

16. Jennifer works 17 hours each week at her part-time job. How many hours does Jennifer work in 4 weeks at her part-time job?

F 42 H 34
G 68 J 51

17. The Pride Club is having a car wash to raise money for a local charity. They charge $3.50 for each car they wash. If they wash 57 cars, how much money will the Pride Club raise?

A $199.50 C $85.50
B $171.50 D $175.00

18. Paco has five $1 bills, three $5 bills, and two $10 bills in his wallet. Without looking, Paco selects one bill from his wallet. What is the probability Paco selected a $1 bill?

F $\frac{1}{5}$ H $\frac{3}{10}$
G $\frac{1}{2}$ J $\frac{1}{8}$

19. Rosita is buying new carpet for her bedroom. Her bedroom is 12 meters by 13 meters. How many square meters of carpet does Rosita need to order for her bedroom?

A 25 C 136
B 156 D 123

20. The distance from Micah's house to school is 7 kilometers. How many meters is this distance?

F 0.7 H 70
G 700 J 7,000

STOP

ANSWER ROW 11 Ⓐ Ⓑ Ⓒ Ⓓ 13 Ⓐ Ⓑ Ⓒ Ⓓ 15 Ⓐ Ⓑ Ⓒ Ⓓ 17 Ⓐ Ⓑ Ⓒ Ⓓ 19 Ⓐ Ⓑ Ⓒ Ⓓ
12 Ⓕ Ⓖ Ⓗ Ⓙ 14 Ⓕ Ⓖ Ⓗ Ⓙ 16 Ⓕ Ⓖ Ⓗ Ⓙ 18 Ⓕ Ⓖ Ⓗ Ⓙ 20 Ⓕ Ⓖ Ⓗ Ⓙ

CHAPTER 9 CUMULATIVE REVIEW

Work each problem.
Find the correct answer.
Mark the space for the answer.

Part 1 Concepts

1. Which statement is true about the remainder in 24)682?
   - A It is equal to 28.
   - B It is less than 5.
   - C It is greater than 12.
   - D It is greater than 9.

2. Which of the following is the dollar amount for seven thousand, five hundred thirty-eight dollars and seventy-two cents?
   - F \$738.72
   - G \$7,538.72
   - H \$7,583.72
   - J \$7,538.27

3. Which of the following represents the probability of an event?
   - A $\frac{\text{number of possible outcomes}}{\text{number of favorable outcomes}}$
   - B $\frac{\text{number of non-favorable outcomes}}{\text{number of favorable outcomes}}$
   - C $\frac{\text{number of favorable outcomes}}{\text{number of possible outcomes}}$
   - D $\frac{\text{number of favorable outcomes}}{10}$

4. What do you multiply by to change 4 feet to inches?
   - F 12
   - G 36
   - H 3
   - J 5,280

Part 2 Computation

5.
$$\begin{array}{r} 487 \\ -129 \\ \hline \end{array}$$
   - A 368
   - B 506
   - C 558
   - D 358

6.
$$\begin{array}{r} 23694 \\ +51918 \\ \hline \end{array}$$
   - F 75,602
   - G 75,502
   - H 75,612
   - J 75,512

7.
$$\begin{array}{r} 569 \\ \times 87 \\ \hline \end{array}$$
   - A 49,503
   - B 48,303
   - C 48,403
   - D 49,403

8. 48)850
   - F 17 r44
   - G 15 r20
   - H 17 r24
   - J 17 r34

9.
$$\begin{array}{r} \$48.62 \\ -17.83 \\ \hline \end{array}$$
   - A \$30.89
   - B \$30.79
   - C \$39.89
   - D \$30.97

10. What is the mean of 78, 64, 79, 63, and 76?
   - F 72
   - G 76
   - H 16
   - J 75

CUMULATIVE REVIEW

ANSWER ROW 1 Ⓐ Ⓑ Ⓒ Ⓓ 3 Ⓐ Ⓑ Ⓒ Ⓓ 5 Ⓐ Ⓑ Ⓒ Ⓓ 7 Ⓐ Ⓑ Ⓒ Ⓓ 9 Ⓐ Ⓑ Ⓒ Ⓓ
2 Ⓕ Ⓖ Ⓗ Ⓙ 4 Ⓕ Ⓖ Ⓗ Ⓙ 6 Ⓕ Ⓖ Ⓗ Ⓙ 8 Ⓕ Ⓖ Ⓗ Ⓙ 10 Ⓕ Ⓖ Ⓗ Ⓙ

11. 9 kg = ________ grams

A 0.009
B 9,000
C 900
D 90

12. 24 feet = ________ yd

F 72
G 6
H 8
J 3

13. What is the area of the figure shown?

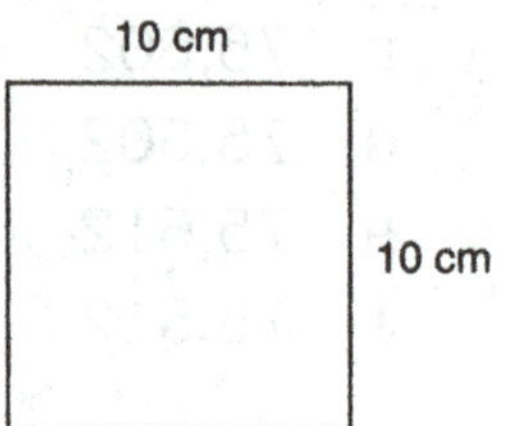

A 10 sq cm
B 100 sq cm
C 20 sq cm
D 40 sq cm

14. What is the perimeter of the figure shown?

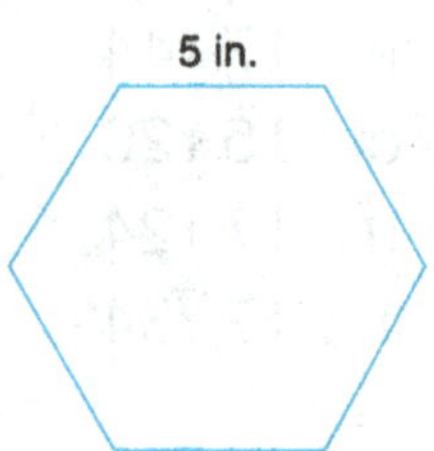

F 5 in.
G 30 in.
H 25 in.
J 35 in.

Part 3 Applications

15. Pablo has 318 coins in his bank. His brother Julian has 467 coins in his bank. How many more coins does Julian have?

A 785
B 159
C 149
D 152

16. For lunch, Cameron bought a sandwich for $2.85, a salad for $1.70, and a fruit juice for $1.12. How much money did Cameron spend on lunch?

F $4.67
G $5.87
H $4.87
J $5.67

17. The pizzeria uses 17 pounds of flour in each batch of dough. How many batches of dough could be made with 1,598 pounds of flour?

A 194
B 78
C 97
D 94

18. During the first seven basketball games, Dante scored the following number of points: 9, 7, 11, 5, 6, 7, and 13. What is the range of the number of points Dante scored in the first seven basketball games?

F 7
G 13
H 8
J 6

19. Bethany's backyard is a rectangle. It is 45 feet by 59 feet. What is the area of Bethany's backyard?

A 208 sq ft
B 2,415 sq ft
C 2,615 sq ft
D 2,655 sq ft

20. Mr. Dean has a rectangular patio in his backyard. It is 6 meters by 4 meters. What is the perimeter of Mr. Dean's patio?

F 20 m
G 24 m
H 16 m
J 28 m

STOP

ANSWER ROW 11 Ⓐ Ⓑ Ⓒ Ⓓ 13 Ⓐ Ⓑ Ⓒ Ⓓ 15 Ⓐ Ⓑ Ⓒ Ⓓ 17 Ⓐ Ⓑ Ⓒ Ⓓ 19 Ⓐ Ⓑ Ⓒ Ⓓ
12 Ⓕ Ⓖ Ⓗ Ⓙ 14 Ⓕ Ⓖ Ⓗ Ⓙ 16 Ⓕ Ⓖ Ⓗ Ⓙ 18 Ⓕ Ⓖ Ⓗ Ⓙ 20 Ⓕ Ⓖ Ⓗ Ⓙ

CHAPTER 10 CUMULATIVE REVIEW

NAME ______________________________

Work each problem.
Find the correct answer.
Mark the space for the answer.

Part 1 Concepts

1. Which problem would you use to find the estimated sum of 576 and 894?

A 500 + 800
B 500 + 900
C 600 + 800
D 600 + 900

2. Which of the following sets of numbers has a median of 26?

F 24, 31, 27, 29, 33
G 34, 26, 22, 27, 25
H 19, 26, 23, 21, 27
J 22, 28, 31, 34, 20

3. Which of the following units of measure would you use to measure the capacity of a bucket?

A meters
B feet
C liters
D grams

4. Which is another way to write $\frac{6}{9}$?

F $\frac{2}{3}$

G $\frac{9}{6}$

H $\frac{12}{15}$

J $\frac{16}{19}$

Part 2 Computation

5. 354×2

A 604 C 608
B 658 D 708

6. $\$94.36 - 7.89$

F $86.47 H $86.48
G $87.57 J $97.57

7. $53\overline{)3233}$

A 68 C 71
B 61 D 53

8. $18\overline{)\$4004.64}$

F $222.47 H $220.45
G $250.01 J $222.48

9. A spinner has four sections of equal size. Two sections are yellow, one section is green, and the other section is blue. What is the probability of the spinner landing on blue?

A $\frac{1}{2}$ C $\frac{1}{3}$

B $\frac{1}{4}$ D $\frac{1}{5}$

10. 70 millimeters = ________ cm

F 7 H 700
G 7,000 J 70,000

CUMULATIVE REVIEW

ANSWER ROW 1 Ⓐ Ⓑ Ⓒ Ⓓ 3 Ⓐ Ⓑ Ⓒ Ⓓ 5 Ⓐ Ⓑ Ⓒ Ⓓ 7 Ⓐ Ⓑ Ⓒ Ⓓ 9 Ⓐ Ⓑ Ⓒ Ⓓ
2 Ⓕ Ⓖ Ⓗ Ⓙ 4 Ⓕ Ⓖ Ⓗ Ⓙ 6 Ⓕ Ⓖ Ⓗ Ⓙ 8 Ⓕ Ⓖ Ⓗ Ⓙ 10 Ⓕ Ⓖ Ⓗ Ⓙ

11. 16 qt = ________ gallons

A 8
C 4
B 2
D 6

12. What is the area of the figure shown?

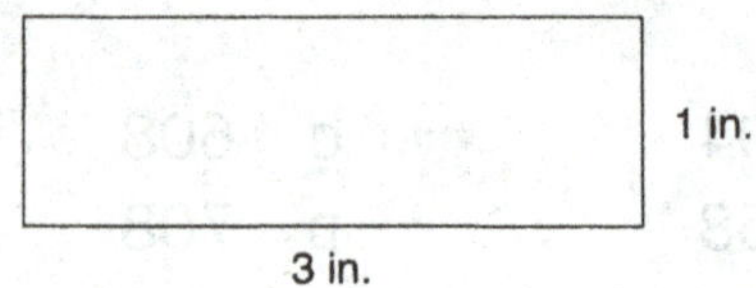

F 1 sq in.
H 6 sq in.
G 3 sq in.
J 8 sq in.

13. What is the greatest common factor of 18 and 32?

A 3
C 6
B 4
D 2

14. Rename $\frac{9}{4}$ as a mixed numeral.

F $1\frac{4}{5}$
H $2\frac{1}{2}$
G $1\frac{1}{2}$
J $2\frac{1}{4}$

Part 3 Applications

15. Dale golfed three games over the weekend. He scored a 67, 59, and a 61. What is the total score of Dale's three golf games?

A 186
C 177
B 187
D 176

16. At the grocery store, turkeys are on sale for $0.99 per pound. Mrs. Stitz bought a 12-pound turkey. How much did the turkey cost?

F $11.76
H $9.99
G $12.00
J $11.88

17. There are 1,326 students at Glenview Middle School. How many classes of 26 students each could there be?

A 36
C 51
B 61
D 26

18. Marta has a 6-pound bowling ball. How many ounces does the bowling ball weigh?

F 72
H 76
G 69
J 96

19. Mr. Carlton built a gazebo in his backyard that has five sides. Each side is 6 feet long. What is the perimeter of Mr. Carlton's gazebo?

A 30 ft
C 50 ft
B 25 ft
D 36 ft

20. Demetre, Todd, Ken, and Dylan ordered a pizza. The pizza was cut into 12 slices. Each of them ate two pieces of pizza. In simplest form, what fraction of the pizza did the boys eat?

F $\frac{1}{12}$
H $\frac{1}{2}$
G $\frac{2}{12}$
J $\frac{2}{3}$

STOP

ANSWER ROW 11 Ⓐ Ⓑ Ⓒ Ⓓ 13 Ⓐ Ⓑ Ⓒ Ⓓ 15 Ⓐ Ⓑ Ⓒ Ⓓ 17 Ⓐ Ⓑ Ⓒ Ⓓ 19 Ⓐ Ⓑ Ⓒ Ⓓ
12 Ⓕ Ⓖ Ⓗ Ⓙ 14 Ⓕ Ⓖ Ⓗ Ⓙ 16 Ⓕ Ⓖ Ⓗ Ⓙ 18 Ⓕ Ⓖ Ⓗ Ⓙ 20 Ⓕ Ⓖ Ⓗ Ⓙ

CHAPTER 11 CUMULATIVE REVIEW

NAME ______________________________

Work each problem.
Find the correct answer.
Mark the space for the answer.

Part 1 Concepts

1. Which of the following quotients is between 48 and 55?
   - A $12\overline{)418}$
   - B $27\overline{)1431}$
   - C $19\overline{)1102}$
   - D $31\overline{)1457}$

2. What are all the factors of 24?
   - F 1, 2, 4, 6, 12, 24
   - G 2, 4, 6, 8, 12
   - H 1, 2, 3, 4, 5, 6, 8, 12, 24
   - J 1, 2, 3, 4, 6, 8, 12, 24

3. Which problem would you use to find the estimated product of 1218 and 578?
   - A 1000 × 500
   - B 1200 × 500
   - C 1000 × 600
   - D 2000 × 600

4. What number is in the denominator of the product of $\frac{3}{7} \times \frac{3}{5}$?
   - F 35
   - G 12
   - H 9
   - J 42

Part 2 Computation

5. 312×42
   - A 13,004
   - B 13,104
   - C 12,104
   - D 12,004

6. $7\overline{)9431}$
   - F 1,347 r1
   - G 1,343
   - H 1,347 r3
   - J 1,347 r2

7. $\$356.70 + \$343.98$
   - A \$699.68
   - B \$700.68
   - C \$700.78
   - D \$699.68

8. What is the mode of 108, 135, 122, 106, 133, 108, and 121?
   - F 119
   - G 29
   - H 108
   - J 121

9. 3,000 mL = ________ liters
   - A 300
   - B 3
   - C 3,000,000
   - D 30

10. 9 cm = ________ mm
    - F 90
    - G 900
    - H 9,000
    - J 9

CUMULATIVE REVIEW

ANSWER ROW 1 Ⓐ Ⓑ Ⓒ Ⓓ 3 Ⓐ Ⓑ Ⓒ Ⓓ 5 Ⓐ Ⓑ Ⓒ Ⓓ 7 Ⓐ Ⓑ Ⓒ Ⓓ 9 Ⓐ Ⓑ Ⓒ Ⓓ
2 Ⓕ Ⓖ Ⓗ Ⓙ 4 Ⓕ Ⓖ Ⓗ Ⓙ 6 Ⓕ Ⓖ Ⓗ Ⓙ 8 Ⓕ Ⓖ Ⓗ Ⓙ 10 Ⓕ Ⓖ Ⓗ Ⓙ

11. 5 yards 1 foot = ________ ft

A 16 C 61
B 72 D 27

12. What is the perimeter of the figure shown?

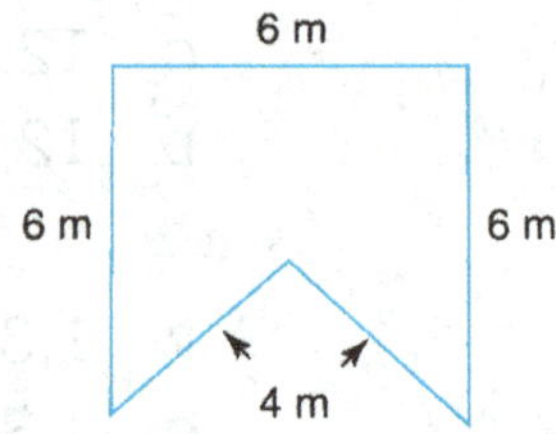

F 22 m H 32 m
G 20 m J 26 m

13. Change $5\frac{6}{7}$ to a fraction.

A $\frac{41}{7}$ C $\frac{30}{7}$
B $\frac{36}{7}$ D $\frac{35}{7}$

14. $\frac{5}{9} \times \frac{6}{10}$

F $\frac{50}{39}$ H $\frac{1}{3}$
G $\frac{3}{5}$ J $\frac{4}{5}$

Part 3 Applications

15. Jacqueline went shopping at the mall. She bought a dress for \$54.83, a pair of shoes for \$21.95, and a necklace for \$17.30. How much money did Jacqueline spend at the mall?

A \$94.08 C \$83.08
B \$93.08 D \$84.08

16. The warehouse inventory shows that there are 5,778 books in the warehouse. There are 18 books in each box. How many boxes of books should there be in the warehouse?

F 321 H 521
G 354 J 368

17. Last week, Silvia kept track of the number of miles she drove each day of the week. She drove the following number of miles Sunday through Saturday: 5, 28, 34, 28, 33, 41, 13. What is the mean number of miles Silvia drove each day last week?

A 26 C 28
B 41 D 182

18. On Saturday, Calvin ran in a 15-kilometer race. How many meters was the race?

F 1,500 H 15,000
G 0.015 J 150

19. On a math quiz, Adrian scored an $\frac{18}{20}$. In simplest form, what was Adrian's score on the math quiz?

A $\frac{6}{10}$ C $\frac{3}{10}$
B $\frac{9}{10}$ D $\frac{1}{2}$

20. Miliya needs 8 ribbons, each one $\frac{3}{4}$ foot long. If she cuts the ribbon from one long ribbon, how long must the ribbon be?

F 8 ft H $6\frac{3}{4}$ ft
G 7 ft J 6 ft

STOP

ANSWER ROW 11 Ⓐ Ⓑ Ⓒ Ⓓ 13 Ⓐ Ⓑ Ⓒ Ⓓ 15 Ⓐ Ⓑ Ⓒ Ⓓ 17 Ⓐ Ⓑ Ⓒ Ⓓ 19 Ⓐ Ⓑ Ⓒ Ⓓ
12 Ⓕ Ⓖ Ⓗ Ⓙ 14 Ⓕ Ⓖ Ⓗ Ⓙ 16 Ⓕ Ⓖ Ⓗ Ⓙ 18 Ⓕ Ⓖ Ⓗ Ⓙ 20 Ⓕ Ⓖ Ⓗ Ⓙ

NAME ______________________

CHAPTER 12 CUMULATIVE REVIEW

Work each problem.
Find the correct answer.
Mark the space for the answer.

Part 1 Concepts

1. In $3,291.54, what number is in the tenths place?

A 1 C 4
B 9 D 5

2. How many digits are in the product of 648 × 92?

F 2 H 5
G 3 J 6

3. When writing $\frac{16}{24}$ in simplest form, by what number do you divide the numerator and denominator?

A the greatest common factor of 16 and 24
B the sum of 16 and 24
C the product of 16 and 24
D the value of the denominator, 24

4. What number is in the numerator of the sum of $\frac{1}{5} + \frac{2}{3}$?

F 13 H 2
G 3 J 15

5. Which number is the divisor in $513\overline{)8461}$?

A 8461 C 16
B 513 D 16 r3

Part 2 Computation

6.

$$\begin{array}{r} 45678 \\ +82902 \\ \hline \end{array}$$

F 128,570
G 129,570
H 129,580
J 128,580

7. $89\overline{)10324}$

A 113
B 116
C 112
D 115

8.

$$\begin{array}{r} \$592.46 \\ -370.95 \\ \hline \end{array}$$

F $963.41
G $222.51
H $863.41
J $221.51

9. 4 lb = ________ ounces

A 16 C 40
B 64 D 48

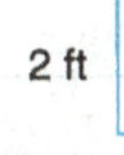

10. What is the area of the figure shown?

F 34 sq ft H 38 sq ft
G 36 sq ft J 40 sq ft

CUMULATIVE REVIEW

ANSWER ROW 1 Ⓐ Ⓑ Ⓒ Ⓓ 3 Ⓐ Ⓑ Ⓒ Ⓓ 5 Ⓐ Ⓑ Ⓒ Ⓓ 7 Ⓐ Ⓑ Ⓒ Ⓓ 9 Ⓐ Ⓑ Ⓒ Ⓓ
2 Ⓕ Ⓖ Ⓗ Ⓙ 4 Ⓕ Ⓖ Ⓗ Ⓙ 6 Ⓕ Ⓖ Ⓗ Ⓙ 8 Ⓕ Ⓖ Ⓗ Ⓙ 10 Ⓕ Ⓖ Ⓗ Ⓙ

11. Write $3\frac{10}{12}$ in simplest form.

A $3\frac{5}{6}$ C $3\frac{5}{12}$

B $3\frac{2}{3}$ D $3\frac{1}{2}$

12. $9 \times \frac{5}{6}$

F $9\frac{4}{5}$

G $9\frac{5}{6}$

H $7\frac{1}{2}$

J $7\frac{1}{3}$

13. $8\frac{3}{16} + 5\frac{7}{8}$

A $14\frac{3}{16}$

B $13\frac{5}{8}$

C $13\frac{1}{16}$

D $14\frac{1}{16}$

Part 3 Applications

Use this graph to answer question 14.

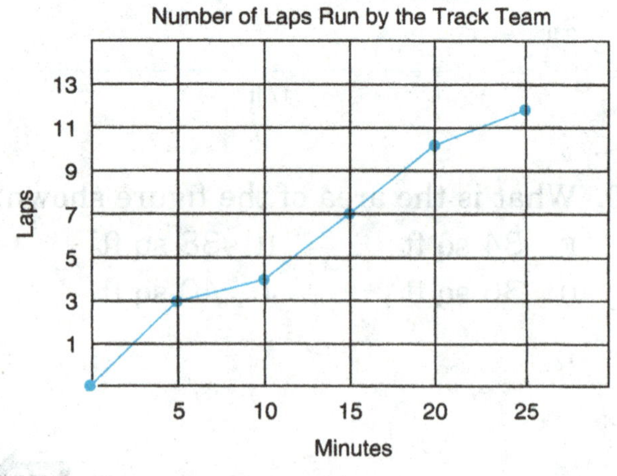

14. How many laps did the track team run in 15 minutes?

F 7 H 8

G 10 J 3

15. Ming's car has a 12-gallon gas tank. A full tank of gas will last for 324 miles. How many miles can Ming's car be driven for each gallon of gas?

A 57 C 54

B 27 D 24

16. Mr. Benton is painting three rooms of his house. He calculated that he will be painting 1,152 square feet. Each can of paint will cover 98 square feet. How many cans of paint will Mr. Benton need to paint all three rooms?

F 111 H 12

G 1,054 J 13

17. A picture frame is $8\frac{1}{2}$ inches long and $5\frac{3}{4}$ inches wide. What is the area of the picture frame?

A $40\frac{3}{8}$ C $41\frac{1}{2}$

B $48\frac{7}{8}$ D $40\frac{1}{2}$

18. Pedro can jump $1\frac{7}{8}$ feet. Maria can jump $1\frac{1}{3}$ feet higher. How many feet can Maria jump?

F $3\frac{1}{8}$ H $3\frac{1}{3}$

G $3\frac{5}{24}$ J $2\frac{7}{8}$

STOP

ANSWER ROW **11** Ⓐ Ⓑ Ⓒ Ⓓ **13** Ⓐ Ⓑ Ⓒ Ⓓ **15** Ⓐ Ⓑ Ⓒ Ⓓ **17** Ⓐ Ⓑ Ⓒ Ⓓ
12 Ⓕ Ⓖ Ⓗ Ⓙ **14** Ⓕ Ⓖ Ⓗ Ⓙ **16** Ⓕ Ⓖ Ⓗ Ⓙ **18** Ⓕ Ⓖ Ⓗ Ⓙ

NAME ______________________

CHAPTER 13 CUMULATIVE REVIEW

Work each problem.
Find the correct answer.
Mark the space for the answer.

Part 1 Concepts

1. Which of the following addition problems has a sum closest to 1,300?

A 964 + 137
B 437 + 981
C 557 + 820
D 615 + 693

2. Which of the following is true?

F 1 pint = 4 cups
G 16 ounces = 1 pound
H 1 foot = 6 inches
J 4,380 feet = 1 mile

3. Which of the following numbers is not prime?

A 7 C 11
B 5 D 9

4. When solving the problem $\frac{13}{16} - \frac{3}{8}$, how should you rename the problem?

F $\frac{13}{16} - \frac{6}{16}$ H $\frac{13}{16} - \frac{3}{16}$

G $\frac{6}{8} - \frac{3}{8}$ J $\frac{13}{16} - \frac{11}{16}$

Part 2 Computation

5. 16 + 9 + 23 + 75

A 103
B 123
C 204
D 120

6. 785 × 546

F 11,775
G 1,331
H 382,980
J 428,610

7. 21)$18.69

A $0.89
B $8.13
C $0.91
D $8.90

8. 6 gal 3 qt = ______ qt

F 24 H 15
G 21 J 27

9. A coin is tossed. What is the probability of it landing on tails?

A 1 C $\frac{1}{2}$
B 2 D $\frac{1}{4}$

10. $\frac{3}{4} \times \frac{7}{8}$

F $\frac{24}{28}$ H $1\frac{5}{8}$
G $\frac{10}{12}$ J $\frac{21}{32}$

CUMULATIVE REVIEW

ANSWER ROW 1 Ⓐ Ⓑ Ⓒ Ⓓ 3 Ⓐ Ⓑ Ⓒ Ⓓ 5 Ⓐ Ⓑ Ⓒ Ⓓ 7 Ⓐ Ⓑ Ⓒ Ⓓ 9 Ⓐ Ⓑ Ⓒ Ⓓ
2 Ⓕ Ⓖ Ⓗ Ⓙ 4 Ⓕ Ⓖ Ⓗ Ⓙ 6 Ⓕ Ⓖ Ⓗ Ⓙ 8 Ⓕ Ⓖ Ⓗ Ⓙ 10 Ⓕ Ⓖ Ⓗ Ⓙ

11. Write $\frac{26}{3}$ as a mixed numeral.

A $8\frac{2}{3}$ C $8\frac{1}{3}$
B $3\frac{1}{3}$ D $9\frac{2}{3}$

12. $\frac{3}{5} \times \frac{4}{9}$

F $\frac{1}{2}$ H $\frac{20}{27}$
G $\frac{6}{7}$ J $\frac{4}{15}$

13. $7\frac{3}{4} + 6\frac{3}{8}$

A $13\frac{3}{4}$
B $14\frac{1}{8}$
C $13\frac{7}{8}$
D $14\frac{3}{4}$

14. $\frac{5}{6} - \frac{1}{9}$

F $\frac{13}{18}$
G $1\frac{1}{3}$
H $\frac{17}{18}$
J $\frac{4}{9}$

Part 3 Applications

15. Benjamin bought a football that cost $14.69. He paid with a $20 bill. How much change did Benjamin get back?

A $34.69 C $6.31
B $5.31 D $16.41

16. Gabrielle works 13 hours a week at her part-time job. How many hours does Gabrielle work in 52 weeks?

F 208 H 65
G 676 J 715

17. A professional basketball court is 94 feet by 50 feet. What is the area of a professional basketball court?

A 4,700 sq ft C 288 sq ft
B 1,584 sq ft D 470 sq ft

18. Shalesha bought $\frac{3}{4}$ pound of turkey. She used $\frac{1}{5}$ of the turkey to make a sandwich. How much turkey did Shalesha use on her sandwich?

F $\frac{4}{9}$ H $\frac{19}{20}$
G $\frac{3}{20}$ J $\frac{11}{20}$

19. Kenyon and Miguel ordered a pizza. Kenyon ate $\frac{1}{4}$ of the pizza. Miguel ate $\frac{1}{3}$ of the pizza. How much of the pizza did Kenyon and Miguel eat?

A $\frac{1}{12}$ C $\frac{2}{7}$
B $\frac{3}{20}$ D $\frac{7}{12}$

20. Meagan has an animal poster and a soccer poster in her room. The animal poster is 4 feet wide and the soccer poster is $2\frac{1}{4}$ feet wide. How much wider is the animal poster?

F $6\frac{14}{4}$ H $1\frac{1}{4}$
G $1\frac{3}{4}$ J $2\frac{3}{4}$

STOP

ANSWER ROW 11 Ⓐ Ⓑ Ⓒ Ⓓ 13 Ⓐ Ⓑ Ⓒ Ⓓ 15 Ⓐ Ⓑ Ⓒ Ⓓ 17 Ⓐ Ⓑ Ⓒ Ⓓ 19 Ⓐ Ⓑ Ⓒ Ⓓ
12 Ⓕ Ⓖ Ⓗ Ⓙ 14 Ⓕ Ⓖ Ⓗ Ⓙ 16 Ⓕ Ⓖ Ⓗ Ⓙ 18 Ⓕ Ⓖ Ⓗ Ⓙ 20 Ⓕ Ⓖ Ⓗ Ⓙ

NAME ______________________

CHAPTER 14 CUMULATIVE REVIEW

Work each problem.
Find the correct answer.
Mark the space for the answer.

Part 1 Concepts

1. What is the estimated quotient of $9\overline{)263}$?

A 20 C 40
B 30 D 55

2. Which of the following numbers is not a factor of 36?

F 3 H 8
G 6 J 12

3. Which of the following fractions is not equivalent to $\frac{12}{18}$?

A $\frac{2}{3}$ C $\frac{4}{6}$
B $\frac{6}{9}$ D $\frac{3}{4}$

4. Which of the following is a quadrilateral?

F pentagon
G triangle
H trapezoid
J octagon

5. Which of the following numbers is prime?

A 7 C 36
B 49 D 15

Part 2 Computation

6. 657×89

F 58,473
G 11,169
H 59,773
J 17,106

7. $40\overline{)2023}$

A 51 r17
B 50 r20
C 102 r23
D 50 r23

8. What is the mean of \$325, \$312, \$367, \$354, and \$312?

F \$334 H \$312
G \$55 J \$325

9. 800 cm = ________ meters

A 80 C 0.8
B 80,000 D 8

10. What is the volume of the figure shown?

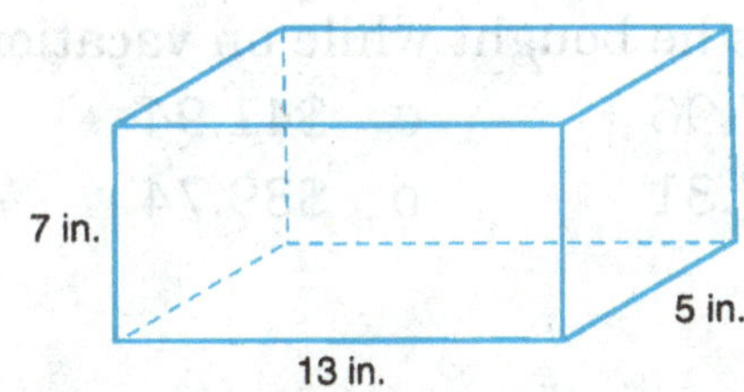

F 150 cubic in. H 5 cubic in.
G 455 cubic in. J 910 cubic in.

CUMULATIVE REVIEW

ANSWER ROW 1 Ⓐ Ⓑ Ⓒ Ⓓ 3 Ⓐ Ⓑ Ⓒ Ⓓ 5 Ⓐ Ⓑ Ⓒ Ⓓ 7 Ⓐ Ⓑ Ⓒ Ⓓ 9 Ⓐ Ⓑ Ⓒ Ⓓ
2 Ⓕ Ⓖ Ⓗ Ⓙ 4 Ⓕ Ⓖ Ⓗ Ⓙ 6 Ⓕ Ⓖ Ⓗ Ⓙ 8 Ⓕ Ⓖ Ⓗ Ⓙ 10 Ⓕ Ⓖ Ⓗ Ⓙ

11. $\frac{1}{3} \times 4\frac{1}{2}$

A $4\frac{1}{6}$ C $1\frac{1}{6}$
B $1\frac{1}{2}$ D $1\frac{2}{3}$

12. $2\frac{1}{9} - \frac{7}{9}$

F $2\frac{2}{9}$
G $1\frac{2}{9}$
H $2\frac{8}{9}$
J $1\frac{1}{3}$

13. Name the figure shown.

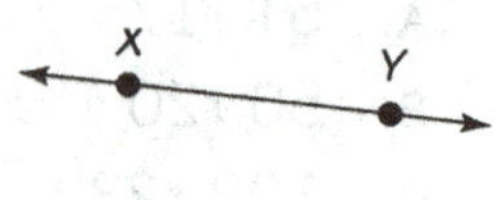

A line segment XY
B line X
C line XY
D line Y

14. Name the figure shown.

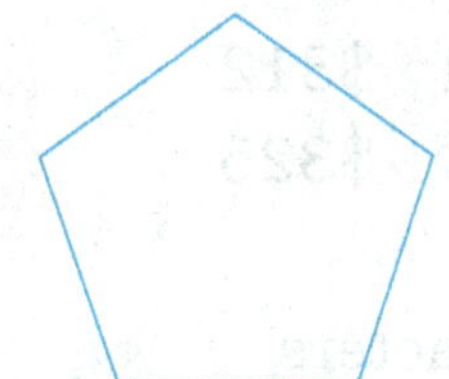

F triangle
G pentagon
H hexagon
J quadrilateral

Part 3 Applications

15. On vacation, Lamar bought three T-shirts. Each T-shirt cost $13.98. How much did Lamar spend on the T-shirts he bought while on vacation?

A $27.96 C $41.94
B $37.31 D $39.74

16. There are 20,736 students attending 32 different schools in a school district. Suppose the same number of students attend each school. How many students attend each school?

F 745 H 597
G 651 J 648

17. Elsa bought 3 pints of fruit juice. How many cups of fruit juice did Elsa buy?

A 6 C 16
B 12 D 9

18. It takes Isaac $5\frac{1}{2}$ hours to drive to his parents' house, and $3\frac{1}{3}$ hours to drive to his grandparents' house. How many more hours does it take Isaac to drive to his parents' house?

F 2 H $2\frac{1}{3}$
G $2\frac{1}{6}$ J $1\frac{1}{6}$

19. Mitch stocks the cereal aisle at the grocery store. Which term best describes the shape of a cereal box?

A circle C rectangular solid
B cube D pentagon

20. Kim read for $1\frac{1}{2}$ hours. Alex read for $\frac{3}{4}$ hour. How much longer did Kim read?

F $\frac{1}{2}$ hour H $\frac{3}{4}$ hour
G $1\frac{1}{2}$ hours J $\frac{1}{4}$ hour

STOP

ANSWER ROW 11 Ⓐ Ⓑ Ⓒ Ⓓ 13 Ⓐ Ⓑ Ⓒ Ⓓ 15 Ⓐ Ⓑ Ⓒ Ⓓ 17 Ⓐ Ⓑ Ⓒ Ⓓ 19 Ⓐ Ⓑ Ⓒ Ⓓ
12 Ⓕ Ⓖ Ⓗ Ⓙ 14 Ⓕ Ⓖ Ⓗ Ⓙ 16 Ⓕ Ⓖ Ⓗ Ⓙ 18 Ⓕ Ⓖ Ⓗ Ⓙ 20 Ⓕ Ⓖ Ⓗ Ⓙ

NAME ______________________

ALGEBRA READINESS
Missing Term—Addition

Subtraction Property of Equality: If you subtract the same number from each side of an equation, the two sides remain equal.

Find the value of the missing term in the equation $x + 4 = 9$ using the Subtraction Property of Equality.

$$x + 4 = 9$$

To undo the addition of 4, subtract 4 from both sides of the equation.

$$x + 4 - 4 = 9 - 4$$
$$x + 0 = 5$$
$$x = 5$$

Ring the operation you should perform on both sides of the equation to find the missing term.

1. $n + 2 = 6$	subtract 6	(subtract 2)	add 2
2. $5 + y = 11$	add 5	subtract 11	subtract 5
3. $x + 8 = 10$	subtract 8	add 8	subtract 10
4. $4 + w = 7$	subtract 7	subtract 4	add 4

Find the value of the missing term in each equation.

5. $x + 3 = 8$ ___5___ $m + 2 = 6$ ______

6. $7 + y = 11$ ______ $4 + d = 7$ ______

7. $n + 2 = 12$ ______ $x + 9 = 14$ ______

8. $8 + w = 15$ ______ $6 + z = 16$ ______

ALGEBRA READINESS

ALGEBRA READINESS
Missing Term—Subtraction

Addition Property of Equality: If you add the same number to each side of an equation, the two sides remain equal.

Find the value of the missing term in the equation $x - 3 = 5$ using the Addition Property of Equality.

$$x - 3 = 5$$

To undo the subtraction of 3, add 3 to both sides of the equation.

$$x - 3 + 3 = 5 + 3$$
$$x + 0 = 8$$
$$x = 8$$

Ring the operation you should perform on both sides of the equation to find the missing term.

1. $n - 4 = 3$	add 3	subtract 4	(add 4)
2. $y - 7 = 2$	add 7	add 2	subtract 2
3. $d - 3 = 8$	subtract 3	add 3	add 8
4. $x - 2 = 11$	add 11	subtract 2	add 2

Find the value of the missing term in each equation.

5. $m - 8 = 2$	10	$d - 4 = 8$	______
6. $x - 3 = 2$	______	$z - 6 = 3$	______
7. $y - 1 = 7$	______	$n - 7 = 7$	______
8. $n - 5 = 6$	______	$x - 2 = 9$	______

NAME ____________________

ALGEBRA READINESS
Missing Term—Multiplication

Division Property of Equality: If you divide each side of an equation by the same nonzero number, the two sides remain equal.

Find the value of the missing term in the equation $4 \times n = 12$ using the Division Property of Equality.

$$4 \times n = 12$$

To undo the multiplication of 4, divide by 4 on both sides of the equation.

$$\frac{4 \times n}{4} = \frac{12}{4}$$

$$n = 3$$

Ring the operation you should perform on both sides of the equation to find the missing term.

1. $7 \times x = 28$	multiply by 7	(divide by 7)	divide by 28
2. $m \times 5 = 25$	divide by 5	multiply by 5	divide by 25
3. $3 \times y = 27$	divide by 27	divide by 3	multiply by 3
4. $n \times 8 = 48$	divide by 48	multiply by 8	divide by 8

Find the value of the missing term in each equation.

5. $3 \times n = 24$ 8 $8 \times w = 32$ ________

6. $w \times 5 = 45$ ________ $m \times 9 = 18$ ________

7. $x \times 7 = 14$ ________ $3 \times d = 33$ ________

8. $6 \times y = 36$ ________ $z \times 7 = 63$ ________

ALGEBRA READINESS

NAME ______________________

ALGEBRA READINESS
Missing Term—Division

Multiplication Property of Equality: If you multiply each side of an equation by the same number, the two sides remain equal.

Find the value of the missing term in the equation $\frac{n}{3} = 6$ using the Multiplication Property of Equality.

$$\frac{n}{3} = 6$$

To undo the division by 3, multiply by 3 on both sides of the equation.

$$\frac{n}{3} \times 3 = 6 \times 3$$

$$n = 18$$

Ring the operation you should perform on both sides of the equation to find the missing term.

1. $\frac{x}{4} = 6$ — (multiply by 4) — divide by 4 — multiply by 6
2. $\frac{a}{2} = 9$ — multiply by 9 — multiply by 2 — divide by 2
3. $\frac{w}{5} = 6$ — divide by 5 — multiply by 6 — multiply by 5
4. $\frac{n}{6} = 8$ — divide by 8 — multiply by 8 — multiply by 6

Find the value of the missing term in each equation.

5. $\frac{n}{2} = 7$ ___14___ $\frac{w}{7} = 8$ ______
6. $\frac{x}{6} = 4$ ______ $\frac{x}{5} = 7$ ______
7. $\frac{m}{8} = 5$ ______ $\frac{n}{12} = 10$ ______
8. $\frac{a}{3} = 12$ ______ $\frac{x}{9} = 8$ ______

NAME ______________________

ALGEBRA READINESS
Mixed Missing Term

Ring the operation you should perform on both sides of the equation to find the missing term.

1. $7 \times n = 49$	multiply by 7	(divide by 7)	divide by 49
2. $a + 6 = 10$	subtract 6	add 6	subtract 10
3. $x - 3 = 8$	subtract 3	add 8	add 3
4. $\frac{w}{4} = 9$	divide by 4	multiply by 4	multiply by 9
5. $8 + y = 13$	subtract 8	subtract 13	add 8
6. $m \times 4 = 48$	multiply by 4	divide by 4	multiply by 48

Find the value of the missing term in each equation.

7. $\frac{w}{6} = 5$ ___30___ $d - 7 = 8$ ________

8. $8 + a = 12$ ________ $\frac{n}{4} = 7$ ________

9. $3 \times x = 21$ ________ $m - 3 = 15$ ________

10. $y - 5 = 4$ ________ $\frac{x}{5} = 12$ ________

11. $9 + n = 16$ ________ $w \times 8 = 64$ ________

ALGEBRA READINESS

NAME ______________________

ALGEBRA READINESS
Function Tables

A **function** is a rule that relates two variables. A function states that for each value of one variable, there is exactly one value related to the other variable.

For example, $y = 4 + x$ is a function. For each value of x, there is exactly one value of y. To find each value of y for a given value of x, add 4 to the given value for x.

A **function table** helps you organize the values of a function. For each entry in the first column, there is exactly one entry in the second column.

Shown is the function table for $y = 4 + x$.

The x-values in the function table are 1, 2, 3, and 4. To find each y-value, add 4 to each x-value.

x	y
1	5
2	6
3	7
4	8

$1 + 4 = 5$
$2 + 4 = 6$
$3 + 4 = 7$
$4 + 4 = 8$

Complete each function table for the given function.

1. $y = 2 + x$

x	y
1	3
2	
3	
4	

$y = 7 + x$

x	y
1	
2	
3	
4	

2. $y = x - 3$

x	y
4	
5	
6	
7	

$y = x - 5$

x	y
10	
15	
20	
25	

Complete each function table for the given function.

3. $y = 2 \times x$

x	y
1	2
2	
3	
4	

$y = 6 \times x$

x	y
1	
2	
3	
4	

4. $y = \frac{x}{3}$

x	y
3	
6	
9	
12	

$y = \frac{x}{5}$

x	y
5	
10	
15	
20	

Create a function table for each function. Use the given values of x in the function table.

5. $y = x + 3; x = 1, 2, 3, 4$

$y = x - 4; x = 5, 6, 7, 8$

6. $y = 3 \times x; x = 1, 2, 3, 4$

$y = \frac{x}{2}; x = 2, 4, 6, 8$

NAME ______________________

ALGEBRA READINESS
Number Patterns

To determine the pattern relating a set of numbers, look at the relationship between each pair of consecutive numbers.

What are the next three numbers in the pattern 1, 4, 7, 10, 13, . . .?

Determine what operation is performed to get from one number to the next number in the pattern.

1, 4, 7, 10, 13, . . .

+3 +3 +3 +3

To get from one number to the next in the pattern, 3 is added.

Add 3 to 13 to get the 6th number in the pattern. $13 + 3 = 16$

Add 3 to 16 to get the 7th number in the pattern. $16 + 3 = 19$

Add 3 to 19 to get the 8th number in the pattern. $19 + 3 = 22$

The the next three numbers in the pattern are 16, 19, 22.

Find the next three numbers of each pattern.

	a		*b*	
1.	2, 7, 12, 17, 22, . . .	27, ______	1, 3, 5, 7, 9, . . .	______
2.	1, 2, 4, 7, 11, 16, . . .	______	84, 78, 72, 66, 60, . . .	______
3.	2, 6, 10, 14, 18, . . .	______	5, 10, 20, 35, 55, . . .	______
4.	108, 96, 84, 72, . . .	______	2, 4, 8, 16, 32, . . .	______
5.	1, 3, 9, 27, . . .	______	320, 160, 80, 40, . . .	______